Engineering Design Process
Second Edition

....

Engineering Design Process
Second Edition

Yousef Haik

University of North Carolina—Greensboro

Tamer Shahin

Kings College London, UK

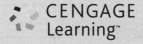

CENGAGE
Learning

Australia • Brazil • Japan • Korea • Mexico • Singapore • Spain • United Kingdom • United States

CENGAGE
Learning™

Engineering Design Process,
Second Edition (International Edition)
Yousef Haik and Tamer Shahin

Publisher, Global Engineering:
Christopher M. Shortt

Senior Acquisitions Editor: Randall Adams

Senior Developmental Editor: Hilda Gowans

Editorial Assistant: Tanya Altieri

Team Assistant: Carly Rizzo

Marketing Manager: Lauren Betsos

Media Editor: Chris Valentine

Content Project Manager: Kelly Hillerich

Production Service: RPK Editorial Services, Inc.

Copyeditor: Shelly Gerger-Knechtl

Proofreader: Martha McMaster

Indexer: Shelly Gerger-Knechtl

Compositor: Integra

Senior Art Director: Michelle Kunkler

Internal Designer: Carmela Pereira

Cover Designer: Andrew Adams

Cover Images: © YAKOBCHUK VASYL/
Shutterstock; © ArchMan/Shutterstock;
© 3DProfi/Shutterstock

Text and Image Permissions Researcher:
Kristiina Paul

Senior Rights Acquisitions Specialist:
Deanna Ettinger

First Print Buyer: Arethea Thomas

Library of Congress Control Number: 2010926399

ISBN-13: 978-0-495-66816-9

ISBN-10: 0-495-66816-8

Cengage Learning
200 First Stamford Place, Suite 400
Stamford, CT 06902
USA

Cengage Learning is a leading provider of customized learning solutions with office locations around the globe, including Singapore, the United Kingdom, Australia, Mexico, Brazil, and Japan. Locate your local office at: **international.cengage.com/region.**

Cengage Learning products are represented in Canada by Nelson Education Ltd.

For your course and learning solutions, visit **www.cengage.com/engineering.**

Purchase any of our products at your local college store or at our preferred online store **www.CengageBrain.com.**

Printed in the United States of America
1 2 3 4 5 6 7 13 12 11 10

To our parents, wives, and children.
To future designers.

Brief Table of Contents

Contents

••

Preface

Design remains the focal point of engineering disciplines; it is what distinguishes engineering from other scientific disciplines. Engineers throughout history have wrestled with problems of water not being where it is needed, of minerals not being close at hand, of building materials having to be moved. Ancient engineers were often called on to devise the means for erecting great monuments, for designing defenses against enemies, and for moving people and goods across rough terrain and even rougher water.

The word engineer originated in the eleventh century and is derived from the Latin origin "ingeniator" meaning one with "ingenium" or the clever one. Before the scientific revolution, ingenuity was demonstrated in many devices. These devices were built by using a simple principle of what works and why it works in this way. Adaptation from nature was prominent in this era. For example, Leonardo da Vinci earned the title *Ingenere* General for his flying device and his bridge design to connect Istanbul to Europe, amongst his many other inventions. Galileo's use of systematic explanation and scientific approach to tackle problems is regarded by historians as the landmark of structured engineering design that is based on scientific merits and mathematical presentation. Following the first Industrial Revolution, beginning in the eighteenth century, the French developed a university engineering education with emphasis on civil engineering, while the British pioneered mechanical engineering. The Industrial Revolution brought a proliferation of new machines and manufacturing techniques and provided an impetus for the growth of science and commerce on an international scale. During the second Industrial Revolution in the middle of nineteenth century, mass production and automation prevailed and were driven by many branches of engineering.

Our modern lifestyle is deeply influenced by our ability to employ scientific discoveries in a wide variety of devices. The continued pursuit of design excellence is empowered by engineers' ability to produce products that meets consumer needs.

In the early 1900s, it was common in American industry for master mechanics to invent and, subsequently, for draftsmen to copy on paper what had been synthesized experimentally in the shop. Since it was less costly and more efficient to erase rather than to remake parts in the shop, the value of synthesizing on paper was soon realized. Recent trends have been to apply theory where appropriate in the process of mechanical design. But overall, it is emphasized that all of the useful ingredients—such as various aspects of art, science, engineering, practical experience, and ingenuity—must be properly blended in the design process. A successful design is achieved when a logical procedure is followed to meet a specific need. This procedure, called the design process, is similar to the scientific method with respect to its step-by-step routine. Often, designs are not accomplished by an engineer simply completing the design steps in the given order. The design process holds within its structure an iterative procedure. As the engineer proceeds through the steps, new information may be discovered and new objectives may be specified, at which time the steps may require revisiting. The more time and effort an engineer spends on articulating the problem definition and understanding the needs statement, the less frequent the need for iteration.

This book is written as an introductory course in design. Students' technical capabilities are assumed to be at the level of college physics and calculus. For students with advanced technical capabilities the analysis part in the design sequence could be emphasized.

This book consists of eleven chapters. Chapter 1 is an overview of the design steps and serves as an introduction to the book. Chapter 2 presents a few design tools that designers must master prior to the design process. Some of these tools serve as an introduction to courses that students will encounter in future course work. Chapters 3 through 9 present the steps of the design process. The author is aware that the sequence of these steps can be changed according to instructor preference. Instructors can alter the presentation sequence without having to change the presentation material. Chapter 10 discusses issues relating to the design cost. Chapter 11 presents a list of project descriptions that can serve as an entry point to instructors' assignments. In this second edition we have integrated design labs with the chapters. The purpose of these labs is to create design activities that help students, especially freshmen and sophomores, to adjust to working in teams. The first few of these labs are geared toward team building. It is anticipated that instructors may want to include other activities in their design classes.

The authors wish to thank all colleagues and students who helped in producing this book, including Dr. Adnan Al-Bashir who provided Lab 5: Project Management. Students are encouraged to submit their comments and suggestions to the authors. The authors also wish to thank the following reviewers for their helpful suggestions: Thomas R. Grimm, Michigan Technological University; Peter Jones, Auburn University; Peter Eliot Weiss, University of Toronto; and Steven C. York, Virginia Tech.

—Yousef Haik and Tamer Shahin

Engineering Design Process
Second Edition

Introduction

One of the first steps in the engineering design process is to have design meetings. In design meetings, engineers, technicians, and other staff members come up with solutions to fill specific customer need. (Zsolt Nyulaszi/Shutterstock)

· ·

1.1 OBJECTIVES

By the end of this chapter, you should be able to

1. Define engineering design.
2. Appreciate the importance and challenges of engineering design.
3. Understand the need for a formalized systematic design process.
4. Name and briefly describe the steps for the design process.
5. Distinguish between different systematic design models.
6. Discuss ethical problems and professional codes of ethics.

1.2 DEFINITION OF ENGINEERING DESIGN

A formal definition of engineering design is found in the curriculum guidelines of the Accreditation Board for Engineering and Technology (ABET). The ABET definition states that engineering design is the process of devising a system, component, or process to meet desired needs. It is a decision-making process (often iterative), in which the basic sciences, mathematics, and engineering sciences are applied to optimally convert resources to meet a stated objective. Among the fundamental elements of the design process is the establishment of objectives and criteria, synthesis, analysis, construction, testing, and evaluation. The engineering design component of a curriculum must include most of the following features: development of student creativity, use of open-ended problems, development and use of modern design theory and methodology, formulation of design problem statement and specifications, production processes, concurrent engineering design, and detailed system description. Furthermore, it is essential to include a variety of realistic constraints, such as economic factors, safety, reliability, aesthetics, ethics, and social impact.

1.2.1 Design Levels

As in any field of human activity, there are different degrees of difficulty. In design, these stages are adaptive design, developed design, and new design.

• *Adaptive design:* In the great majority of instances, the designer's work will be concerned with the adaptation of existing designs. There are branches of manufacturing in which development has practically ceased, so that there is hardly anything left for the designer to do except make minor modifications, usually in the dimensions of the product. Design activity of this kind demands no special knowledge or skill, and the problems presented are easily solved by a designer with ordinary technical training. One such example can be the elevator, which has remained the same technically and conceptually for some time now. Another example is a washing machine. This has been based on the same conceptual design for the last several years and varies in only a few parameters, such as its dimensions, materials, and detailed power specifications.

- *Development design:* Considerably more scientific training and design ability are needed for development design. The designer starts from an existing design, but the final outcome may differ markedly from the initial product. Examples of this development could be from a manual gearbox in a car to an automatic one and from the traditional tube-based television to the modern plasma and LCD versions.
- *New design:* Only a small number of designs are new designs. This is possibly the most difficult level in that generating a new concept involves mastering all the previous skills in addition to creativity and imagination, insight, and foresight. Examples of this are the design of the first automobile, airplane, or even the wheel (a long time ago). Try to think of entirely new designs which have been introduced over the last decade.

1.3 IMPORTANCE AND CHALLENGES OF ENGINEERING DESIGN

From the definition in the previous section, it is evident that design is both a scientific and a creative process. Albert Einstein asserted that imagination is more important than knowledge, for knowledge is finite whereas imagination is infinite.

It is essential to realize that design does not start with an engineering drawing made on a computer package such as Pro/Engineer™ or even AutoCAD™. Such a final engineering drawing can be regarded in some ways as the 'lab report' of your final design and hence a method of communicating your design with other people. There are many steps before this, and these steps will be discussed throughout the course of this chapter and in more detail in the rest of the book. Of course, as computer packages become more advanced, designers are able to start using them earlier on in the design process to aid them with their design. However, as design is a creative process, most of the input will have to come from the designer.

Design is widely regarded as one of the most important steps in the development of a product. Indeed, without a design, there would be no product! Not only this, but no matter how good the manufacturing, production, sales, etc. are, if a product is poorly designed, the end product still will be a bad idea and will ultimately fail, as no one likes to purchase a bad idea.

Most consumers will not be aware or even interested in the detailed technical specifications of a product or how efficient the manufacturing process. The first thing that a consumer will usually look at before deciding to purchase something is its design and 'how it looks'. This will be followed by the reliability and quality of the item, then by the price. Think about how people choose to buy a coffee machine or even a mobile phone.

It is interesting to note here that price does not always come first. Many people are willing to pay a bit more if they see the benefits, and this is usually reflected in the design. In some cases, people will only purchase an item if it is expensive for reassurance of quality and possibly prestige (no one would buy a Rolls Royce or a Rolex simply because it was cheap!). However, in most cases, part of the design process will be to design for minimum cost so that a product can be competitive in the marketplace.

Many sources, including the United Kingdom Department of Trade and Industry (DTI) identified that investing money and resources at the design stage yields the biggest return on investment of a product. One of the reasons for this is that changes can be made easily at this early stage, whereas later on, changes in the manufacturing methods and so on could be extremely costly—both in time and money.

Throughout history, humans have been successfully designing artifacts to satisfy the needs of civilization. History is full of great designs and inventions. Recently, design has been driven to meet an existing requirement, to reduce a hazard or an inconvenience, or to develop a new approach. Not all that engineers build has become successful; occasionally, catastrophic failures occur. A few of the well-publicized disasters associated with engineering systems are as follows:

- The Chernobyl nuclear power plant disaster occurred in 1996. According to the World Health Organization (WHO), this lead to the evacuation and resettlement of over 336,000 people, 56 direct deaths, 4000 thyroid cancer cases among children, and approximately 6.6 million people highly exposed to radiation.
- The *Challenger* space shuttle exploded in 1986 after an O-ring seal in its right solid-rocket booster failed. This caused a flame leak, which reached the external fuel tank. The space shuttle was destroyed in 73 seconds after takeoff, and all crew members died.
- The loss of the cabin roof during the flight of a Boeing 737 in 1988 caused one crew member to be blown out of the airplane. Age and the design of the aircraft, which relied on stress to be alleviated by controlled breakaway zones, were ultimately to blame.
- A skywalk at the Kansas City Hyatt Regency Hotel collapsed just after the hotel was opened in 1981. The skywalk rods were not designed to hold the combined weights of the walkways and the 2000 people that had gathered on the them. 200 people were injured, and 114 were killed.
- A crack in an engine pylon caused the loss of an engine and the subsequent crash of a DC-10 airplane in 1979, killing 273 people.
- The design layout of the fuel tanks was the cause of the Concorde crash in 2000, killing 113 people. When the aircraft struck debris on the runway, the tire that subsequently exploded caused a tank to rupture. The Concorde's airworthiness certificate was revoked, and all Concorde airplanes remained grounded for 15 months. This eventually contributed to the demise of supersonic passenger planes.
- The crash of Columbia Space Shuttle in 2003 was attributed to the detachment of a piece of debris from the external tank bipod attach region and striking the underside or leading edge of the port wing of the Columbia.

Walton lists the reason for failures in most engineering designs:

- Incorrect or overextended assumptions
- Poor understanding of the problem to be solved
- Incorrect design specifications
- Faulty manufacturing and assembly
- Error in design calculations
- Incomplete experimentation and inadequate data collection
- Errors in drawings
- Faulty reasoning from good assumptions

As can be seen, all of the disaster examples given and the reasons for their catastrophic failures can be summarized and categorized within one or more of the items Walton lists.

However, even if a design is a technical success and no faults occur, many designs still fail to achieve their desired goals, and many achieve them but are not adopted by the users. So why do many people fail at design? One of the answers is that design is inherently difficult and a major challenge. Designers not only have to have the creative and technical skills to develop an idea to become a reality, but they also need to predict the future in some ways. They need to predict each step of the product's life from visualization to realization and finally to the end of its life cycle and how it will be disposed of and/or recycled. This means that a designer needs to develop a product that sponsors will like and fund (and so on and so forth) all the way down to the distributors, vendors, users, operators, and society as a whole.

To complicate matters further, everyone has a different opinion/desire on how a product should be designed. A pair of identical twins brought up in the same environment easily can walk into a shop where one can pick up a mobile phone and say it was the most beautiful thing he has ever seen, while the other may decide that it is ghastly. This makes predicting whether people will like and use a product developed by the designer somewhat of a challenge.

It is for these reasons that the systematic design process was introduced to help guide the designer to achieve his/her goals without hindering creativity. The following section discusses this in more detail.

1.4 INTRODUCTION TO SYSTEMATIC DESIGN

Engineering students during their training are presented with a vast amount of theoretical material and information. They only realize their weakness when they are faced with the task of logically applying what they have learned to a specific end. As long as their work is based on familiar models or previous designs, the knowledge they possess is perfectly adequate to enable them to find a solution along conventional lines. As soon as they are required to develop something already in existence to a more advanced stage or to create something entirely new without a previous design, they will fail miserably, unless they have reached a higher level of understanding. Without a set of guidelines, they are at a loss for a starting point and a clear finishing goal line. The design process was formalized to enable both students and professional designers to follow a systematic approach to design and help them guide their creativity and technical problem-solving skills to a satisfactory end.

There are various forms of the systematic design process, and different people list as few as four steps to as many as nine. Essentially though, they all revolve around the same following basic principles:

- Requirements
- Product concept
- Solution concept
- Embodiment design
- Detailed design

The most important step of the design process is identifying the needs of the customer or the 'Requirements' stage. However, before this is done, it is important to establish who the customers are. A vital concept to grasp here is that customers are not only the end users. Customers of a product are everyone who will deal with the product at some stage during its lifetime. For example, the person who will sell the product is also a customer. A designer must make the product attractive for the seller to agree to advertise and market it. Another example of a customer is the person who will service and maintain the product during its lifetime in operation. If a product is difficult to maintain and/or service, independent service providers will be keen to recommend other products or charge more to service the item. And so on. Let us take a look at possible customers of an airplane. These can include:

- Passengers
- Crew
- Pilot
- Airport
- Engineers and service crew
- Fueling companies
- Airlines
- Manufacturing and production departments
- Baggage handlers
- Cleaning and catering companies
- Sales and marketing
- Accounts and finance departments
- Military/Courier/Cargo/etc
- Authorities and official bodies
- Companies involved with the items that will be outsourced

Each of these customers has entirely different (and sometimes conflicting) needs for the same product, and by identifying these customers first, it is then possible to identify all the needs and arrive at a reasonable compromise according to priority and feasibility.

Most of the time the customer provides a generic statement of need, and it is up to the engineer to identify the specific needs of the customer. For example, we are required to design a chair that can be used by a child. Clearly, all of us know how to sit on a chair, so in that perspective, we know how a chair works. A chair is used for sitting. Unfortunately, this description does not say how the chair is made. What material is used? Is the chair flexible or rigid? Does the chair rotate or is it fixed? What does it mean that the chair is to be used by a child? Is safety the biggest concern? How much will the chair cost? How old is the child? And so on.

Many factors in engineering design are not based on a mathematical model, but the engineering design process is maintained to be systematic. In the previous example, we can use Newton's laws of equilibrium to describe the forces generated in the chair's legs. We can also describe the deformation of the legs when a person sits down. We can even use finite element analysis to estimate the stresses in the legs, the seat, and the back. We can also describe the manufacturing process and the joints used in the chair. Some of these factors can be represented using mathematical models, but what mathematical model will describe the color of the chair, and what mathematical model will measure if the chair is safe for use by a child? The function of the chair cannot be completely presented by a mathematical model. These functions are manipulated by reasoning. Different schemes may be needed to describe a certain design (e.g., analytical models identify the geometric presentation of the artifact, economic schemes describe the cost of producing the artifact, verbal contents describe the function of the artifacts, and so on). In the designing process, mathematical modeling, although important, is not sufficient to design the artifact.

The design engineer must learn to think independently, to draw conclusions, and to combine solutions. Many believe that they can acquire this skill set by attending lectures and reading textbooks. They fail to realize that they are only accumulating one fresh item of knowledge after another. Understanding, logical deduction, and judgment cannot be conferred from outside; on the contrary, they are acquired only by diligent thinking and working with the knowledge already possessed. A basic precondition for independent design is a lively imagination. Such imagination is required to do original work.

1.5 DESIGN PROCESS

To design is to create a new product that turns into profit and benefits society in some way. The design process is a sequence of events and a set of guidelines that helps define a clear starting point that takes the designer from visualizing a product in his/her imagination to realizing it in real life in a systematic manner—without hindering their creative process.

The ability to design requires both science and art. The science can be learned through a systematic process (outlined in this chapter), experience, and problem-solving technique (all of which will be mastered during your college education). The art is gained by practice and a total dedication to becoming proficient.

The design of a device or system can be done in one of two ways:

1. *Evolutionary change*: A product is allowed to evolve over a period of time with only slight improvement. This is done when there is no competition. The creative capabilities of the designer are limited.
2. *Innovation:* Rapid scientific growth and technological discoveries as well as competition among companies for their slice of the market have placed a great deal of emphasis on new products, which draw heavily on innovation. The creative skills and analytical ability of the design engineer play an important role.

The invention of the telephone was a truly innovative design. Since its invention, many then tried to evolve and hence improve it over many decades, but very little actually changed until the next innovative and technological jump occurred, and that was the mobile phone. This created a whole new market along with new competition, and since then, this technology has been evolving once more—every once in a while showing signs of new innovation, such as the inclusion of cameras and video-calling and the integration of pda, internet access, and mp3 facilities into one device. Proficient designers control evolution and innovation so they occur simultaneously. Although the emphasis is on innovation, designers must test their ideas against prior design. Engineers can design for the future but must base results on the past.

Section 1.4 summarized the principle steps of a systematic design process and mentioned that various people have slightly different forms of representing such a process. This section describes these different steps in more detail and introduces a selection of various viewpoints from different authors. Figures 1.1 through 1.3 show some of these in chart form.

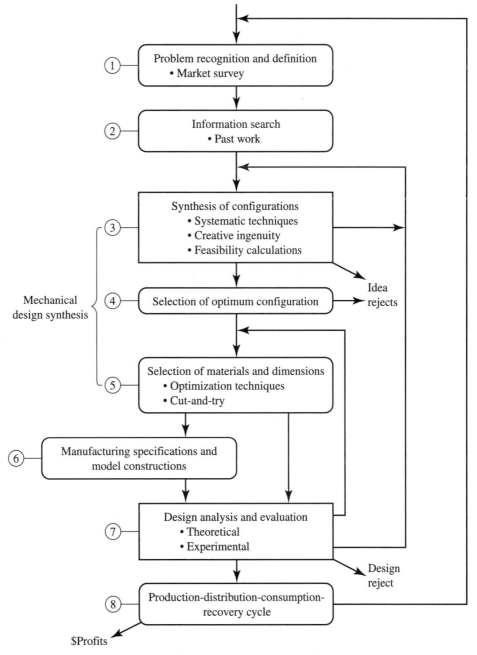

Figure 1.1   Design process map. (From MECHANICAL DESIGN SYNTHESIS, 2/e by Ray C. Johnson. Copyright © 1978. Reprinted by permission of Krieger Publishing Company.)

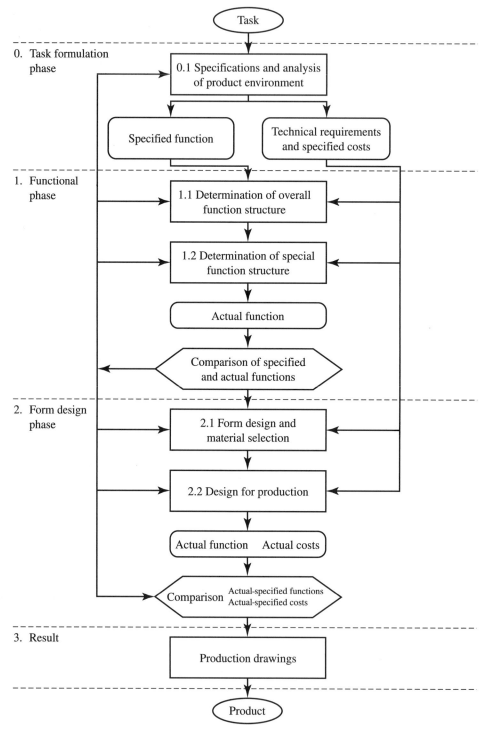

Figure 1.2   Design process map. ("Design Process Map" from ENGINEERING DESIGN: A SYNTHESIS OF VIEWS by C.L. Dym. Copyright © 1994. Reprinted with the permission of Cambridge University Press.)

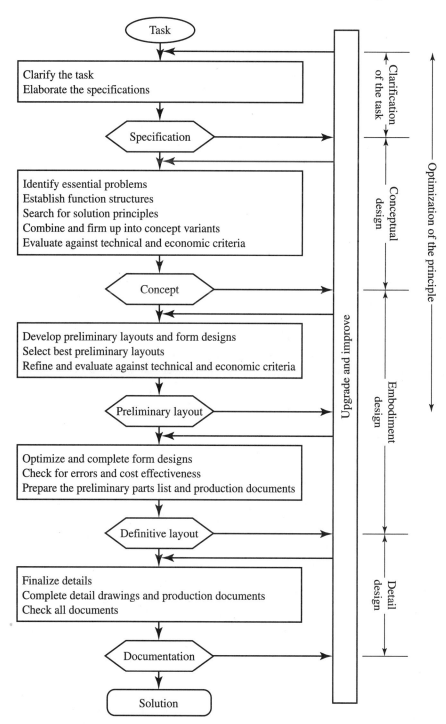

Figure 1.3 Design process map. (From ENGINEERING DESIGN: A SYSTEMATIC APPROACH by G. Pahl and W. Beitz, translated by Ken Wallace, Lucienne Blessing and Frank Bauert, Edited by Ken Wallace. Copyright © Springer-Verlag London Limited 1996. Reprinted by permission.)

The difference in these charts is in sequence names. Careful examination of the charts leads to identification of virtually the same stages. Some stages are combined in one process. However, closer inspection of the timeline of these charts hints towards a trend to further formalizing the design process and leaning more towards addressing the problem and postpone the solution to the latter stages rather than finding a solution early on and then trying to improve it. This is most apparent between Figure 1.1, which was published in 1978, and Figure 1.2, which was published in 1994. Figure 1.3 is widely regarded as the benchmark for the modern systematic design process, and new representations are invariably a modified version of this chart. This is also true of Figure 1.4, which illustrates the design steps that are adopted in this book. It also represents a roadmap to the book on how it is structured and reference to the relevant chapters is to be found within the figure. The steps demonstrated are iterative and require a series of decisions to move the design along. More often a design oscillates back and forth between stages until it reaches an acceptable form and can move to the next stage. A brief description of these stages is summarized in the following subsections.

1.5.1 Identifying Customer Needs (Requirements)

The need for a new design can be generated from several sources, including the following:

- *Client request:* In a design company, a client may submit a request for developing an artifact. It is often unlikely that the need will be expressed clearly. The client may know only the type of product that he or she wants; for example, "I need a safe ladder."

- *Modification of an existing design:* Often a client asks for a modification of an existing artifact to make it simpler and easy to use. In addition, companies may want to provide customers with new, easy-to-use products. For example, in a market search you may notice many brand names for coffee makers and the differences among them, such as shape, material used, cost, or special features. As another example, Figures 1.5 through 1.8 demonstrate design developments for paperclips. Each of these designs has its own advantages over the other clips. For example, the endless filament paperclip can be used from either side of the clip. One may argue that the different designs are based on the human evolution of designs and birth of new ideas; however, the major driving force for the renovation of designs is to keep companies in business. The first patent for a paperclip was filed in 1899, and the latest was filed in 1994—a hundred years of paperclip development and innovations.

- *Generation of a new product:* In all profit-oriented industries, the attention, talent, and abilities of management, engineering, production, inspection, advertising, marketing, sales, and servicing are focused on causing the product to return profit for the company and in turn for company stockholders. Unfortunately, sooner or later, every product is preempted by another or degenerates into profitless price competition. For an industry to survive in today's world, it must continue to grow; it cannot afford to remain static. This growth, throughout history, has been built on new products. New products have a characteristic lifecycle pattern in sales volume and profit margins, as shown in Figure 1.9. A product will peak out when it has saturated the market and then begin to decline. It is obvious that an industry must seek out and promote a flow of new product ideas. These new products are usually protected by applying for patents.

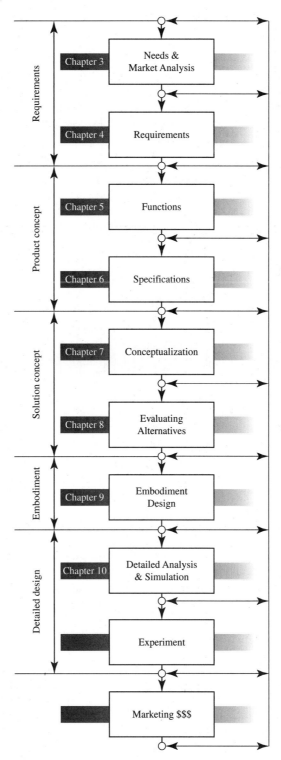

Figure 1.4 Design process.

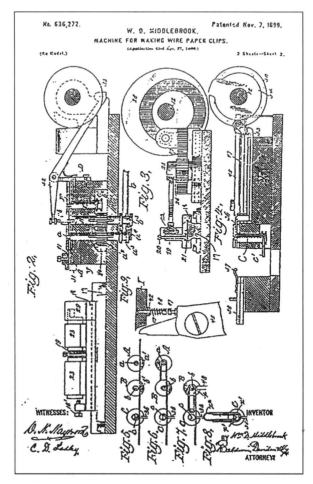

Figure 1.5 William Middelbrook patent for a machine for making paperclips (1899).

1.5.2 **Market Analysis (Requirements)**

Designers must locate what is already available in the market and what they have to offer. Information gathering is a vital task. Some companies hire design engineers so they can get away from paying royalties to patent holders. Design engineers may consult the following sources to determine market availability:

- Technical and trade journals
- Abstracts
- Research reports
- Technical libraries
- Catalog of component suppliers
- U.S. Patent Office
- The Internet

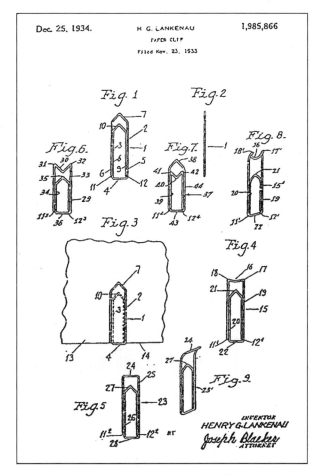

Figure 1.6   Patent for Gothic-style paperclip, issued in 1934.

The information gathered may reveal an available design solution and the hardware to accomplish the goal. Sometimes, the goal may be altered to produce a requested product or abandoned if the product already exists. Knowledge of existing products will save the designer and client time and money. Once the designer determines what is in the market, creativity should be directed towards generating alternatives. Chapter 3 discusses Market analysis and the gathering of information in more detail.

1.5.3 Defining Goals (Requirements)

In this stage of the design process, the designer defines what must be done to resolve the need(s). The definition is a general statement of the desired end product. Many of the difficulties encountered in design may be traced to poorly stated goals, or goals that were hastily written and resulted in confusion or too much flexibility.

In almost all cases, the client request comes in a vague verbal statement such as, "I need an aluminum can crusher." or "I need a safe ladder." Designers must recognize that

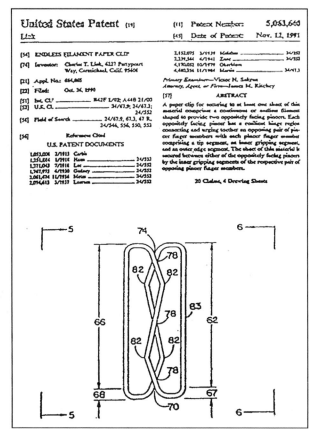

Figure 1.7   Patent for an endless filament paper clip, issued to Charles T. Link (1991).

customer needs are not the same as product specifications. Needs should be expressed in functional terms. Customers will offer solutions; designers must determine the real needs, define the problem, and act accordingly. During the customer interview, the designer must listen carefully to what the customer has to say. The designer's function is to clarify the client's design requirements. An *objective tree* may be constructed for clarification.

Often the need statement and goals are combined into one process. An objective tree is a tool used by designers to organize the customer's wants into categories; Chapter 4 discusses both *needs* and *goals* in more detail.

1.5.4 Establishing Functions (Product Concept)

Recognizing the generality of the need statement and where the problem/need stands in the whole system is a fundamental element in the design process. There is a big difference between being asked to design a car suspension system and designing a car. It is useful to consider the level at which the designer is asked to work. It is also useful to identify the functions instead of the potential solutions. This is sometimes referred to as

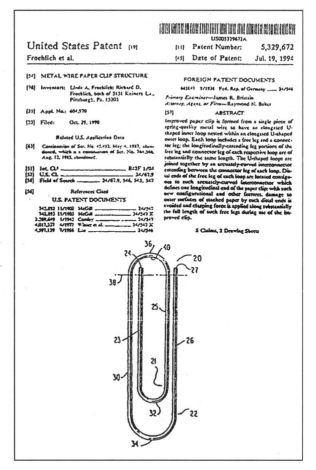

Figure 1.8 Patent for a large paperclip made of spring wire, issued to Linda and Richard Froehlich (1994).

remaining 'solution neutral' (i.e., no solution is referred to at this stage). In reality, the designer is trying to assess what actions the product should perform during its lifetime and operation. This technique allows for alternatives to be explored that can address the needs and goals rather than fixating on a solution that the customer provides unintentionally early on. For example, a client may ask for a traffic light system to be placed on a particular junction, where in fact, an underpass may be a more viable solution to achieve the real goals of the task, which may be to alleviate traffic congestion. This stage of the design process demonstrates one of the advantages of systematic design in that it guides the designer to a problem-focused design rather than a solution-focused one. Another example: A blood bank approached a designer to find a solution for its frequently broken centrifuge. When the centrifuge breaks down, the blood separation unit shuts down. The blood bank has tried replacing the centrifuge, but after a few months, it breaks down. The need statement is to fix the centrifuge in such a way that it will

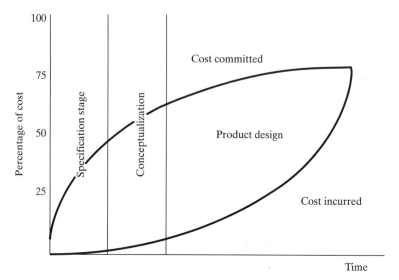

Figure 1.9   Basic lifecycle of new product.

reduce the break-down time. A designer who recognizes the design process will not jump on fixing the centrifuge; he will ask the following question: "What is the function we try to accomplish with the centrifuge?" The answer is to separate blood by enhancing the gravitational pull. Then the function is to separate blood cells from the whole blood. This clear statement enables the designer to find alternative solutions other than the centrifuge. Often the functions will be divided into subfunctions, and they will define the requirements of the artifact.

A further example: If you were required to design a lawn mower, the function should be defined as a method for shortening the grass. The difference between two definitions of a lawn mower (designing a lawn mower or method to shorten grass) is that one may limit your imagination and creativity while the other may give you free reign for creativity. Chapter 5 discusses establishing functional structure in more detail.

1.5.5 Task Specifications (Product Concept)

This stage requires the designer to list all pertinent data and parameters that tend to control the design and guide it towards the desired goal; it also sets limits on the acceptable solutions. It should not be defined too narrowly, because the designer will eliminate acceptable solutions. However, it cannot be too broad or vague, because this will leave the designer with no direction to satisfy the design goal. Let's reconsider the lawn mower example and set some task specifications.

1. It should be safe to operate, especially when used near children.
2. It should be easy to operate by an average person.
3. The device should be powered (either manually or by other means).
4. The device should be easily stored in a garage.

5. The device should allow combinations of lawn tasks:
   a. Collecting leaves and mowing the lawn.
   b. Chopping the leaves and fertilizing.
   c. Chopping branches for mulch.
6. The device should use available hardware and require sheet metal, fiber glass, or plastic.
7. Material should be selected based on cost, manufacturability, strength, appearance, and ability to withstand varying weather conditions.
8. The device must be reliable and not require frequent maintenance.
9. The selling price should be lower than that in the market.

1.5.6 Conceptualization (Solution Concept)

The process of generating alternative solutions to the stated goal in the form of concepts requires creative ability. The conceptualization starts with generating new ideas. In this stage, the designer must review the market analysis and the task specifications as he or she engages in the process of innovation and creativity. This activity usually requires *free-hand sketches* for producing a series of alternative solutions. The alternatives do not need to be worked out in detail but are recorded as possibilities to be tested. Alternatives to perform the functions should be listed in an organized fashion. For example, to design a novel transportation system, a designer may list the methods as follows.

1. Natural way
   a. Human
      i. Walk
      ii. Swim
   b. Animal
      i. Ride
      ii. Pulled in a cart
2. With aids
   a. Land
      i. Bike
      ii. Skate
   b. Water
      i. Canoe
      ii. Tube
   c. Air
      i. Kite
   d. Mechanical
      i. Land
         • Car
         • Train
         • Tube
      ii. Water
         • Ship
         • Sled

        iii. Air
- Plane
- Rocket

Chapter 7 covers this stage of the design process in more detail

1.5.7 Evaluating Alternatives (Solution Concept)

Once a number of concepts have been generated in sufficient detail, a decision must be made about which one or ones will enter the next, most expensive, stages of the design process. An excellent technique to guide the designer in making the best decision regarding these alternatives is a *scoring matrix*, which forces a more penetrating study of each alternative against specified criteria. Chapter 8 covers this stage of the design process in more detail.

1.5.8 Embodiment Design

Once the concept has been finalized, the next stage is known as the embodiment design, and this is where the product that is being designed begins to take shape. This stage does not include any details yet (no dimensions or tolerances, etc.) but will begin to illustrate a clear definition of a part, how it will look, and how it interfaces with the rest of the parts in the product assembly. This stage is separated from both the conceptual design and the detailed design in that new technologies can replace old ones based on the exact same concept. For example, The concept of a traffic light system will remain the same (three lights: red, amber, and green), perform the same functions and specifications, and work conceptually the same way, but as technologies advance, the lights themselves can change from bulb to LEDs or the way the lights change can be from using a timer to cycle through the lights to using a system that is connected to a modern traffic network. Possibly, the future may hold a system where the traffic light is able to sense the most efficient light for the junction to alleviate congestion and change the lights accordingly. The concept still remains the same, but the execution and parts or the 'embodiment' of the design can change. Chapter 9 discusses embodiment design in more detail.

1.5.9 Analysis and Optimization

Once a possible solution for the stated goal has been chosen, the synthesis phase of the design has been completed and the analysis phase begins. This is also known as 'Detailed Design' and is what most of the engineering courses in an undergraduate degree program cover. In essence, the solution must be tested against the physical laws. The manufacturability of the chosen product also must be checked to ensure its usefulness. A product may satisfy the physical laws, but if it cannot be manufactured, it is a useless product. This stage is put in iterative sequencing with the original synthesis phase. Often, analysis requires a concept to be altered or redefined then reanalyzed, so that the design is constantly shifted between analysis and synthesis. Analysis starts with estimation and is followed by order of magnitude calculation.

    Estimation is an educated guess based on experience. Order of magnitude analysis is a rough calculation of the specified problem. The order of magnitude does not provide an exact solution, but it gives the order in which the solution should be expected. Chapter 10 covers some of the aspects you will need to perform. However, as detailed design covers the majority of an undergraduate engineering program and differs from product to product, details of this stage are beyond the scope of this book.

1.5.10 Experiment

The experiment stage in engineering design requires that a piece of hardware is constructed and tested to verify the concept and analysis of the design as to its work ability, durability, and performance characteristics. Here the design on paper is transformed into a physical reality. Three techniques of construction are available to the designer:

1. *Mock-up:* The mock-up is generally constructed to scale from plastics, wood, cardboard, and so forth. The mock-up is often used to check clearance, assembly technique, manufacturing considerations, and appearance. It is the least expensive technique, provides the least amount of information, and is quick and relatively easy to build.

2. *Model:* This is a representation of the physical system through a mathematical similitude. Four types of models are used to predict behavior of the real system:
   a. A true model is an exact geometric reproduction of the real system, built to scale, and satisfying all restrictions imposed in the design parameters.
   b. An adequate model is so constructed to test specific characteristics of the design.
   c. A distorted model purposely violates one or more design conditions. This violation is often required when it is difficult to satisfy the specified conditions.
   d. Dissimilar models bear no apparent resemblance to the real system, but through appropriate analogies, they give accurate information on behavioral characteristics.

3. *Prototype:* This is the most expensive experimental technique and the one producing the greatest amount of useful information. The prototype is the constructed, full-scale working physical system. Here the designer sees his or her idea come to life and learns about such things as appropriate construction techniques, assembly procedures, work ability, durability, and performance under actual environmental conditions.

As a general rule, when entering the experimental stage of the design process, one should first deal with the mock-up, then the model, and finally the prototype (after the mock-up and model have proven the real worth of the design), to allow beneficial interaction with concept and analysis. Chapter 10 covers this section in more detail.

1.5.11 Marketing

This stage requires specific information that defines the device, system, or process. Here the designer is required to put his or her thoughts regarding the design on paper for the purpose of communication with others. Communication is involved in selling the idea to management or the client, directing the shop on how to construct the design, and serving management in the initial stages of commercialization.

The description should take the form of one of the following:

1. A report containing: a detailed description of the device, how it satisfies the need and how it works, a detailed assembly drawing, specifications for construction, a list of standard parts, a cost breakdown, and any other information that will ensure that the design will be understood and constructed exactly as the designer intended.
2. A flyer containing a list of the special features that the design can provide, advertisements, promotional literature, market testing, and so forth.

Although this section is predominantly beyond the scope of this book, it is possible to refer to Chapters 2 and 3, which cover some of the essential skills needed to succeed here.

1.6 PROFESSIONALISM AND ETHICS

Before you delve into the rest of this book, it is important to understand how you, as a student, will develop into a professional engineer. This will dictate the way in which you deal with all your professional issues. Understanding the concept of being a professional engineer may sound easy, but adopting it and living by it involves far more depth and effort.

Professionalism is a way of life. A professional person is one who engages in an activity that requires a specialized and comprehensive education and is motivated by a strong desire to serve humanity. The work of engineers generally affects the day-to-day life of all humans. Developing a professional frame of mind begins with your engineering education. Being a professional should imply that in addition to providing the specialized work that is expected of him/her, the professional engineer should provide such services with honesty, integrity, and morality. In this spirit, many have tried to be more explicit and have developed rules and guidelines with which to adhere to. Providing a set of rules to be followed in all circumstances is not as straightforward as it might seem as some of the rules will be problem-, profession-, and situation-dependent. Inevitably, others have discussed the ethical implications of the profession as a set of moral values that are associated with culture and religion. With a subject like this, debates will continue, and there will possibly never be a definitive set of rules that the entire world can agree on. However, several engineering societies have developed a code of ethics that must be followed by its member engineers. This serves as an acceptable compromise, and from time to time, these codes get reviewed and updated when necessary.

In such a small introduction, it is immediately apparent that ethics is a complex and still emerging subject. Because of its importance, the authors have designed a lab that deals with ethics (Lab 1: Ethics) and presents several case studies for examination, discussion, and debate. Students should do this lab immediately after covering this section. For additional reading, several online resources deal with ethics, including.

- http://web.mit.edu/ethics/www/essays/probcase.html
- http://onlineethics.org

1.6.1 NSPE Code of Ethics

As mentioned in the previous section, many engineering societies now include their version of a code of ethics by which their members must adhere to. This book refers to the code of ethics by the National Society of Professional Engineers (NSPE).[1] It is updated from time to time, and this section refers to the January 2006 *Code of Ethics*. Other engineering societies have a very similar code of ethics, which will vary in style or wording, but they all invariably require that engineers uphold and advance the integrity, honor, and dignity of the engineering profession by

- Using their knowledge and skill for the enhancement of human welfare.
- Being honest and impartial, and serving with fidelity the public, their employers and clients.
- Striving to increase the competence and prestige of the engineering profession.

The NSPE Code of Ethics is divided into three main sections:

1. **The Fundamental Canons:** These are the main issues that govern a professional engineer from an ethical and professional standing.
2. **Rules of Practice:** This section discusses the first five points of the fundamental canons in more detail.
3. **Professional Obligations:** This section discusses the last point of the fundamental canons in more detail and is focused towards professional conduct from a legal, ethical and societical viewpoint.

The remaining part of this chapter lists excerpts from the NSPE *Code of Ethics*. Students should read and discuss the points raised and try to think of examples whereby they could be applied in the workplace.

The Fundamental Canons

While fulfilling their professional duties, engineers shall

1. Hold paramount the safety, health, and welfare of the public.
2. Perform services only in areas of their competence.
3. Issue public statements only in an objective and truthful manner.
4. Act for each employer or client as faithful agents or trustees.
5. Avoid deceptive acts.
6. Conduct themselves honorably, responsibly, ethically, and lawfully so as to enhance the honor, reputation, and usefulness of the profession.

Rules of Practice

1. Engineers shall hold paramount the safety, health, and welfare of the public.
   a. If engineers' judgment is overruled under circumstances that endanger life or property, they shall notify their employer or client and such other authority as may be appropriate.

---

[1]Reprinted by Permission of the National Society of Professional Engineers (NSPE) www.nspe.org.

    b. Engineers shall approve only those engineering documents that are in conformity with applicable standards.

    c. Engineers shall not reveal facts, data, or information without the prior consent of the client or employer except as authorized or required by law or this Code.

    d. Engineers shall not permit the use of their name or associate in business ventures with any person or firm that they believe is engaged in fraudulent or dishonest enterprise.

    e. Engineers shall not aid or abet the unlawful practice of engineering by a person or firm.

    f. Engineers having knowledge of any alleged violation of this Code shall report thereon to appropriate professional bodies and, when relevant, also to public authorities, and cooperate with the proper authorities in furnishing such information or assistance as may be required.

2. Engineers shall perform services only in the areas of their competence.

    a. Engineers shall undertake assignments only when qualified by education or experience in the specific technical fields involved.

    b. Engineers shall not affix their signatures to any plans or documents dealing with subject matter in which they lack competence, nor to any plan or document not prepared under their direction and control.

    c. Engineers may accept assignments and assume responsibility for coordination of an entire project and sign and seal the engineering documents for the entire project, provided that each technical segment is signed and sealed only by the qualified engineers who prepared the segment.

3. Engineers shall issue public statements only in an objective and truthful manner.

    a. Engineers shall be objective and truthful in professional reports, statements, or testimony. They shall include all relevant and pertinent information in such reports, statements, or testimony, which should bear the date indicating when it was current.

    b. Engineers may express publicly technical opinions that are founded upon knowledge of the facts and competence in the subject matter.

    c. Engineers shall issue no statements, criticisms, or arguments on technical matters that are inspired or paid for by interested parties, unless they have prefaced their comments by explicitly identifying the interested parties on whose behalf they are speaking, and by revealing the existence of any interest the engineers may have in the matters.

4. Engineers shall act for each employer or client as faithful agents or trustees.

    a. Engineers shall disclose all known or potential conflicts of interest that could influence or appear to influence their judgment or the quality of their services.

    b. Engineers shall not accept compensation, financial or otherwise, from more than one party for services on the same project, or for services pertaining to the same project, unless the circumstances are fully disclosed and agreed to by all interested parties.

    c. Engineers shall not solicit or accept financial or other valuable consideration, directly or indirectly, from outside agents in connection with the work for which they are responsible.

    d. Engineers in public service as members, advisors, or employees of a governmental or quasi-governmental body or department shall not participate in decisions with respect to services solicited or provided by them or their organizations in private or public engineering practice.

    e. Engineers shall not solicit or accept a contract from a governmental body on which a principal or officer of their organization serves as a member.

5. Engineers shall avoid deceptive acts.

    a. Engineers shall not falsify their qualifications or permit misrepresentation of their or their associates' qualifications. They shall not misrepresent or exaggerate their responsibility in or for the subject matter of prior assignments. Brochures or other presentations incident to the solicitation of employment shall not misrepresent pertinent facts concerning employers, employees, associates, joint venturers, or past accomplishments.

    b. Engineers shall not offer, give, solicit, or receive, either directly or indirectly, any contribution to influence the award of a contract by public authority, or which may be reasonably construed by the public as having the effect or intent of influencing the awarding of a contract. They shall not offer any gift or other valuable consideration in order to secure work. They shall not pay a commission, percentage, or brokerage fee in order to secure work, except to a bona fide employee or bona fide established commercial or marketing agencies retained by them.

Professional Obligations

1. Engineers shall be guided in all their relations by the highest standards of honesty and integrity.

    a. Engineers shall acknowledge their errors and shall not distort or alter the facts.

    b. Engineers shall advise their clients or employers when they believe a project will not be successful.

    c. Engineers shall not accept outside employment to the detriment of their regular work or interest. Before accepting any outside engineering employment, they will notify their employers.

    d. Engineers shall not attempt to attract an engineer from another employer by false or misleading pretenses.

    e. Engineers shall not promote their own interest at the expense of the dignity and integrity of the profession.

2. Engineers shall at all times strive to serve the public interest.

    a. Engineers shall seek opportunities to participate in civic affairs; career guidance for youths; and work for the advancement of the safety, health, and well-being of their community.

    b. Engineers shall not complete, sign, or seal plans and/or specifications that are not in conformity with applicable engineering standards. If the client or employer insists on such unprofessional conduct, they shall notify the proper authorities and withdraw from further service on the project.

    c. Engineers shall endeavor to extend public knowledge and appreciation of engineering and its achievements.

    d. Engineers shall strive to adhere to the principles of sustainable development in order to protect the environment for future generations.

3. Engineers shall avoid all conduct or practice that deceives the public.
   a. Engineers shall avoid the use of statements containing a material misrepresentation of fact or omitting a material fact.
   b. Consistent with the foregoing, engineers may advertise for recruitment of personnel.
   c. Consistent with the foregoing, engineers may prepare articles for the lay or technical press, but such articles shall not imply credit to the author for work performed by others.
4. Engineers shall not disclose, without consent, confidential information concerning the business affairs or technical processes of any present or former client, employer, or public body on which they serve.
   a. Engineers shall not, without the consent of all interested parties, promote or arrange for new employment or practice in connection with a specific project for which the engineer has gained particular and specialized knowledge.
   b. Engineers shall not, without the consent of all interested parties, participate in or represent an adversary interest in connection with a specific project or proceeding in which the engineer has gained particular specialized knowledge on behalf of a former client or employer.
5. Engineers shall not be influenced in their professional duties by conflicting interests.
   a. Engineers shall not accept financial or other considerations, including free engineering designs, from material or equipment suppliers for specifying their product.
   b. Engineers shall not accept commissions or allowances, directly or indirectly, from contractors or other parties dealing with clients or employers of the engineer in connection with work for which the engineer is responsible.
6. Engineers shall not attempt to obtain employment or advancement or professional engagements by untruthfully criticizing other engineers, or by other improper or questionable methods.
   a. Engineers shall not request, propose, or accept a commission on a contingent basis under circumstances in which their judgment may be compromised.
   b. Engineers in salaried positions shall accept part-time engineering work only to the extent consistent with policies of the employer and in accordance with ethical considerations.
   c. Engineers shall not, without consent, use equipment, supplies, laboratory, or office facilities of an employer to carry on outside private practice.
7. Engineers shall not attempt to injure, maliciously or falsely, directly or indirectly, the professional reputation, prospects, practice, or employment of other engineers. Engineers who believe others are guilty of unethical or illegal practice shall present such information to the proper authority for action.
   a. Engineers in private practice shall not review the work of another engineer for the same client, except with the knowledge of such engineer or unless the connection of such engineer with the work has been terminated.
   b. Engineers in governmental, industrial, or educational employ are entitled to review and evaluate the work of other engineers when so required by their employment duties.
   c. Engineers in sales or industrial employ are entitled to make engineering comparisons of represented products with products of other suppliers.

8. Engineers shall accept personal responsibility for their professional activities, provided, however, that engineers may seek indemnification for services arising out of their practice for other than gross negligence, where the engineer's interests cannot otherwise be protected.
   a. Engineers shall conform with state registration laws in the practice of engineering.
   b. Engineers shall not use association with a nonengineer, a corporation, or partnership as a "cloak" for unethical acts.
9. Engineers shall give credit for engineering work to those to whom credit is due and will recognize the proprietary interests of others.
   a. Engineers shall, whenever possible, name the person or persons who may be individually responsible for designs, inventions, writings, or other accomplishments.
   b. Engineers using designs supplied by a client recognize that the designs remain the property of the client and may not be duplicated by the engineer for others without express permission.
   c. Engineers—before undertaking work for others in connection with which the engineer may make improvements, plans, designs, inventions, or other records that may justify copyrights or patents—should enter into a positive agreement regarding ownership.
   d. Engineers' designs, data, records, and notes referring exclusively to an employer's work are the employer's property. The employer should indemnify the engineer for use of the information for any purpose other than the original purpose.
   e. Engineers shall continue their professional development throughout their careers and should keep current in their specialty fields by engaging in professional practice, participating in continuing education courses, reading in the technical literature, and attending professional meetings and seminars.

LAB 1: Ethics

Ethics is a part of all professional careers, but play an extremely important role in engineering. In this lab, we will discuss the importance of engineering ethics using two case studies.

Purpose

Ethics is a part of all professional careers. This lab introduces case scenarios that deal with professional ethics. Before you start this lab you will need to visit the following Web pages and familiarize yourself with their content:

1. http://web.mit.edu/ethics/www/essays/probcase.html by Caroline Whitebeck.
2. http://www.cwru.edu/affil/wwwethics/. After you review the content of this page, click on Problems. This will lead you to different case studies in engineering and science ethics.
3. http://www.onlineethics.org.

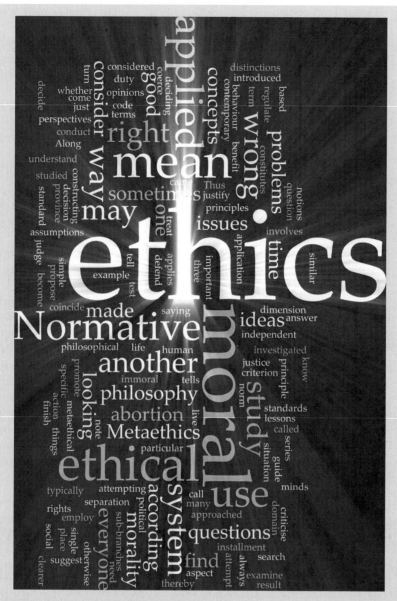

Kheng Guan Toh/Shutterstock

Procedure

You will need to submit two reports in the course of this lab.

1. *Report I assignment:* In this report you should (as a group)
   a. List the major points that were discussed by the Caroline Whitebeck paper.
   b. Discuss the paper and point out where you agree or disagree.
   c. Discuss your scenario and report your findings and arguments (The instructor will assign an ethical scenario to your team.)

2. *Report II assignment*
    a. Identify experts from the university and local community.
    b. Interview at least three experts and report your interview discussion.
    c. Analyze the three interviews.
    d. Make a recommendation/final argument based on your opinion and the interviews.
    e. Create a Web page for your report.

The following are the different scenarios that were obtained from the Internet. The discussion at end of each scenario will help you get started; do not limit your scope to those questions and discussion points when you report.

Scenario I[2]

You are an engineer charged with performing safety testing and obtaining appropriate regulatory agency or outside testing laboratory ("agency") approvals of your company's product. The Gee-Whiz Mark 2 (GWM2) has been tested and found compliant with both voluntary and mandatory safety standards in North America and Europe. Because of a purchase-order error and subsequent oversights in manufacture, 25,000 units of GWM2 ("bad units") were built that are not compliant with any of the North American or European safety standards. A user would be much more vulnerable to electric shock from a bad unit than from a compliant unit. Under some plausible combinations of events, users of the bad unit could be electrocuted.

Retrofitting these products to make them compliant is not feasible, because the rework costs would exceed the profit margin by far. All agree that, because of this defect, the agency safety labels will not be attached to the bad units, as per the requirements of the several agencies. Only two options exist:

1. Scrap the units and take the loss.
2. Sell the units.

An employee of the company notes that many countries have no safety standards of any kind for this type of product. It is suggested that the bad units be marketed in these countries. It is pointed out that many of these nations have no electrical wiring codes; if codes exist, they are not enforced. The argument is thus advanced that the bad GWM2 units are no worse than the modus operandi of the electrical practice of these countries. Assume that no treaties or export regulations would be violated in marketing the bad units to these countries.

Discussion

1. What is your recommendation?
2. Suppose one of the countries under consideration was the country of origin for you or your recent ancestors. Would this affect your recommendation?
3. Now suppose you are not asked for a recommendation, only an opinion. What is your response?
4. Suppose it is suggested that the bad units be sold to a third party, who would very likely sell the units to these countries. What is your comment?

---

[2]Scenarios in Business and Engineering Settings by Joseph H. Wujek and Deborah G. Johnson, from http://www.onlineethics.org/cms/7335.aspx. Used by permission.

5. You are offered gratis one of the bad units for your use at home, provided that you sign a release indicating your awareness of the condition of the unit and that it is given to you as a test unit. (Assume that you can't retrofit it, and that the product could be very useful to you.) Would you accept the offer?

6. Suppose it is suggested that the offer in item 5 be made to all employees of the company. Your comment?

Scenario II[3]

The United States Federal Communications Commission (FCC) Rule Part 15J applies to virtually every digital device (with a few exceptions) manufactured in the United States. The manufacturer must test and certify that the equipment does not exceed FCC-mandated limits for the generation of communications interference caused by conducted and radiated emissions. The certification consists of a report sent to the FCC for review. It is largely an honor system, because the FCC has only a small staff to review an enormous number of applications. The FCC then issues a label ID to be attached to each unit that authorizes marketing of the product. Prior to receiving the label the manufacturer cannot offer for sale or advertise the product. An EMC (electromagnetic compatibility) consultant operating a test site installs a new antenna system and finds that it results in E-field measurements consistently higher than those obtained with the old antennas. Both track within the site-calibration limits, and both antenna vendors claim National Institute of Standards and Technology (formerly National Bureau of Standards) traceability. Which system is the better in absolute calibration is thus unknown. There is not enough time to resolve this discrepancy before a client's new product must be tested for FCC Rules Part 15J compliance.

Discussion

1. Which antenna system should be used to test the product?
2. Is averaging the results ethical, assuming that engineering judgment indicates that this procedure is valid?
3. Suppose the site was never properly (scientifically or statistically) calibrated. Should this fact be made known voluntarily to the FCC?

Scenario III[4]

You are an engineer working in a manufacturing facility that uses toxic chemicals in processing. Your job has nothing to do with the use and control of these materials.

The chemical MegaX is used at the site. Recent stories in the news have reported alleged immediate and long-term human genetic hazards from inhalation or other contact with the chemical. The news items are based on findings from laboratory experiments, done on mice, by a graduate student at a well-respected university physiology department. Other scientists have neither confirmed nor refuted the experimental findings. Federal and local governments have not made official pronouncements on the subject. Several employee friends have approached you on the subject and asked you to do something to eliminate the use of MegaX at your factory. You mention this concern to your manager, who tells you, "Don't worry, we have an industrial safety specialist who handles that." Two months pass, and MegaX is still

[3,4]Scenarios in Business and Engineering Settings by Joseph H. Wujek and Deborah G. Johnson, from http://www.onlineethics.org/cms/7335.aspx. Used by permission.

used in the factory. The controversy in the press continues, but there is no further scientific evidence (pro or con) in the matter. The use of the chemical in your plant has increased, and now more workers are exposed daily to the substance than was the case two months ago.

Discussion

1. What, if anything, do you do?
2. Suppose you again mention the matter to your manager and are told, "Forget it, it's not your job." What should you do now?
3. Your sister works with the chemical. What is your advice to her?
4. Your pregnant sister works with the chemical. What is your advice to her?
5. The company announces a voluntary phasing out of the chemical over the next two years. What is your reaction to this?
6. A person representing a local political activist group approaches you and asks you to make available to them company information regarding the amounts of MegaX in use at the factory and the conditions of use. Do you comply? Why or why not?

Scenario IV[5]

The Zilch Materials Corporation employs you as a test engineer. The company recently introduced a new two-component composition-resin casting material, Megazilch, which is believed to have been well tested by the company and a few selected potential customers. All test results prior to committing to production indicated that the material meets all published specifications and is superior in performance and lower in estimated cost than competitors' materials used in the same kinds of applications.

Potential and committed applications for Megazilch include such diverse products as infants' toys, office equipment parts, interior furnishings of commercial aircraft, and the case material for many electronic products. Marketing estimates predict a 25% increase in the corporation's revenues in the first year after the product is shipped in production quantities.

The product is already in production and many shipments have been made when you discover, to your horror, that under some conditions of storage temperature and other (as yet) unknown factors, the shelf life of the product is seriously degraded. In particular, it will no longer meet specifications for flame retardation if stored for more than 60 days before mixing, instead of the 24 months stated in the published specifications. Its tensile and compressive strengths are reduced significantly as well.

Substantial quantities have been shipped, and the age and temperature history of the lots shipped are not traceable. To recall these would involve great financial loss and embarrassment to the company, and at this point, it is not clear that the shelf life can be improved. Only you and a subordinate, a competent test technician, know of the problem.

Assume that no quick fixes by chemical or physical means are possible, and that the problem is real. That is, there are no mistakes in the scientific findings.

---

[5]Scenarios in Business and Engineering Settings by Joseph H. Wujek and Deborah G. Johnson, from http://www.onlineethics.org/cms/7335.aspx. Used by permission.

Discussion

1. What is the first action you would take relevant to this matter?
2. Suppose you express concern to your immediate supervisor, who tells you, "Forget it! It's no big deal, and we can correct it later. Let me handle this."
3. Suppose further that you detect no action after several weeks have passed since you told your supervisor. What now?
4. In item 3, assume you speak to your supervisor, who then tells you, "I spoke to the executive staff about it and they concur. We'll keep shipping product and work hard to fix it. We've already taken out all the stops, people are working very hard to correct the problem." What, if anything, do you do?
5. It is now three months since you told your supervisor, and in a test of product sampled from current shipments, you see that no fix has been incorporated. What now?

Scenario V[6]

Marsha is employed as the City Engineer by the city of Oz, which has requested bids for equipment to be installed in a public facility. Oz is bound by law to purchase the lowest bid that meets the procurement specifications except "for cause." The low bidder, by a very narrow margin, is Diogenes Industries, a local company. The Diogenes proposal meets the specifications. Marsha recommends purchase of the equipment from Diogenes.

After the equipment is installed, it is discovered that John, the Chief Engineer for Diogenes, is the spouse of Marsha. John was the engineer who had charge of the proposal to Oz, including the final authority on setting the price. As a result of this, Marsha is requested to resign her position for breach of the public trust.

Discussion

1. Was the city justified in seeking Marsha's termination of employment?
2. Suppose Marsha had never been asked to sign a conflict of interest statement. Would this affect your response to question 1?
3. Given the conditions of question 2, suppose Marsha had mentioned, before going to bid, in casual conversation with other persons involved in the procurement that she was married to the Chief Engineer at Diogenes. Does this affect your response?
4. Suppose Marsha and John were not married but shared a household. Does this affect your response?
5. Now suppose Marsha had made known officially her relation to John and the potential for conflict of interest before soliciting bids. Then suppose Marsha rejects the Diogenes bid because she is concerned about the appearance of conflict of interest. She then recommends purchase of the next lowest bid, which meets the specifications. Comment on Marsha's action. ■

---

[6]Scenarios in Business and Engineering Settings by Joseph H. Wujek and Deborah G. Johnson, from http://www.onlineethics.org/cms/7335.aspx. Used by permission.

LAB 2: Ethics and Moral Frameworks

This lab will take you through the process needed to approach and resolve ethical dilemmas that can and will present themselves many times during a design project and your professional career as an engineer. It is important to note that, although there are certain scenarios that are clear cut 'ethical' or 'unethical', there are other times when this line is not so clear. It is beyond the scope of this book to delve into too much depth, and there are other books that cover this vast area sufficiently. Nevertheless, the aim of this lab is to give you a foundation and a feel in order to enable you to identify, discuss, and provide an ethical resolution for ethical issues.

Theory—Code of Ethics and Moral Frameworks

Although strongly related to each other, there is a distinct difference between the terms 'ethics' and 'morals.' In most cases, both are needed together to make a well-balanced judgment. Ethics relates to the philosophy behind a moral outcome and determines the working of a social system. These are usually presented as a set of rules that dictate right or wrong behavior. The NSPE Code of Ethics is one such example and is presented in the main body of Chapter 1. The Code of Ethics alone, though, does not cover everything and at times individual canons can conflict with each other. For example, fundamental canon 4 in the NSPE code of ethics states: "Act for each employer or client as faithful agents or trustees." This dictates that you should remain faithful to your employer. However, for example, in some cases, your employer may be deceiving the public and informing them that the product they are selling performs a certain function, whereas in actual fact it does not. This will be in direct conflict with fundamental canon 5: "Avoid Deceptive Acts." In this case, what do you do?

Morals, on the other hand, define our character and usually address 'appropriate' and 'expected' behavior. Morals deal with adopted codes of conduct or frameworks within a given environment, conception, and/or time. Such codes can deal with controversial behavior, prohibitions, standards of belief systems, and social conformity. Moral frameworks can be abstract and the same outcome can be deemed 'appropriate' or 'inappropriate', depending on the situation. For example, people can argue that "murder is immoral" but during war and on the battlefield that "murder is acceptable."

In some cases, moral frameworks are too abstract to point to a conclusive ethical resolution. Parallel to this, ethics can be regarded as an application of morality. This is why both are usually needed together to form a balanced opinion. There are five main moral frameworks or approaches and they are briefly introduced here.

The Utilitarian Approach (Utilitarianism)

Some ethicists emphasize that ethical action is the one that provides the most good, does the least harm, or (put another way) produces the greatest balance of good over harm. The ethical corporate action then is the one that produces the greatest good and does the least harm for all who are affected—customers, employees, shareholders, the community, and the environment. The utilitarian approach deals with consequences; it tries both to increase the good done and to reduce the harm done. A typical example usually discussed with this approach is given here.

You are hiking alone in a forest and you come up to a village where there is a terrorist holding 20 people hostage. The terrorist is about to kill all 20 people, but somehow you convince him not to kill anyone. He agrees as long as you take the gun and kill one of them to prove his point (whatever that may be). If you choose not to kill one person, then he will proceed to kill all 20. There are no other options in this situation. What would you do? Discuss this within your group.

The Rights Approach

This approach aims to make decisions based on actions that best protect and respect the moral rights of those affected. It begins with the belief that humans have a dignity based on their human nature as well as the ability to choose freely what they do with their lives. On the basis of such dignity, they have a right to be treated as ends and not merely as means to other ends. The list of moral rights includes the rights to make one's own choices about what kind of life to lead, to be told the truth, not to be injured, to have a degree of privacy, and so on is widely debated. Recently, moral development has progressed towards providing rights for non-humans as well.

The Fairness or Justice Approach

This approach is based on contributions from Aristotle and other Greek philosophers and revolves around the idea that all equals should be treated equally. This also suggests that it may be fair to treat all that are unequal unequally. We pay people more based on their harder work or the greater amount that they contribute to an organization and say that is fair. But what about CEO salaries that are hundreds of times larger than the pay of others? Many ask whether the huge disparity is based on a defensible standard or whether it is the result of an imbalance of power and hence is unfair.

The Common Good Approach

Greek philosophers also have contributed the notion that life in a community is good in itself, and our actions should contribute to that life. This approach suggests that the interlocking relationships of society are the basis of ethical reasoning and that respect and compassion for all others—especially the vulnerable—are requirements of such reasoning. This approach also calls attention to the common conditions that are important to the welfare of everyone. This may be a system of laws, effective police and fire departments, health care, a public educational system, or even public recreational areas.

The Virtue Approach

A very ancient approach to ethics is that ethical actions ought to be consistent with certain ideal virtues that provide for the full development of our humanity. These virtues are dispositions and habits that enable us to act according to the highest potential of our character and on behalf of values like truth and beauty. Honesty, courage, compassion, generosity, tolerance, love, fidelity, integrity, fairness, self-control, and prudence are all examples of virtues. Virtues ask of any action, "What kind of person will I become if I do this?" or "Is this action consistent with my acting at my best?"

Putting the Approaches Together

Each of these approaches helps us determine what standards of behavior can be considered ethical. Although seemingly straightforward, it unfortunately is not so simple. The first problem is that we may not agree on the content of some of these specific approaches. We may not all agree to the same set of human and civil rights. We may not agree on what constitutes the common good. We may not even agree on what is good and what is harmful.

The second problem is that the different approaches may not all answer the question "What is ethical?" in the same way. Nonetheless, each approach gives us important information with which to determine what is ethical in a particular circumstance. More often than not, the different approaches do lead to similar answers.

Moral Reasoning and Approaching Ethical Dilemmas

Now that you are aware of both the Code of Ethics and moral frameworks, you can try to tackle some ethical problems. To do that, you will need to follow a process such as the one listed here. The more you practice, the more natural this process will become, and ultimately, this should become an integral and habitual aspect of your profession.

Step 1. *Identify the Ethical Dilemma*
   - Will any of your options be damaging to someone, some group, or some particular thing? Do the potential decisions involve a choice between a good and bad alternative, or between two 'goods' or even two 'bads?'
   - What are the relevant facts of the case? What facts are not known? Try to establish a timeline if relevant. Can you learn more about the situation? Do you know enough to make a decision?
   - Who are the involved parties? What individuals and groups have an important stake in the outcome? Are some concerns more important? Why?

Step 2. *Identify the Relevant Codes of Ethics*
   - List all of the relevant codes of ethics that affect both the problems identified in step 1 as well as the codes relevant to any potential decisions you may make.
   - List the possible solutions that you may take based on the codes of ethics alone.

Step 3. *Establish any Possible Potential Conflicts within the Identified Relevant Codes of Ethics*
   - Do any of the identified codes conflict with each other? Group them together.
   - Is there one which is more important to address over the other? Why?

Step 4. *Evaluate the Moral Frameworks to Resolve any Conflicting Codes of Ethics*
   - Which option will produce the most good and do the least harm? (The Utilitarian Approach)
   - Which option best respects the rights of all who have a stake? (The Rights Approach)
   - Which option treats people equally or proportionately? (The Justice Approach)
   - Which option best serves the community as a whole, not just some members? (The Common Good Approach)
   - Which option leads you to act as the sort of person you want to be? (The Virtue Approach)

Step 5. *Make a Recommendation/Decision*
   - Considering the Code of Ethics and the given moral frameworks, which option best addresses the situation? You should structure your recommendation as a well thought out argument based on a model such as the one provided by Toulmin [1] (explained next).
   - Ask yourself: If you told someone you respect or announced to the public the recommendation you have chosen, what would they say?
   - How can your decision be implemented with the greatest care and attention to the concerns of all stakeholders?

Toulmin's Model for Argumentation and Moral Reasoning

Delving further into these problem solving steps, you need to be able to reason and argue (both to yourself and to others) the decision you will make and justify your choices. Stephen Toulmin [1] broke down the structure of an argument into a logical model that includes all of the elements necessary to complete an argument from start to finish. These points should be addressed each time you make a 'claim' (as described next) at any stage of the five steps described previously. The model also should be used to present your final argument and decision.

Essential Elements of an Argument:

1. *Claim*   This is as the word suggests. This is your claim or statement that has no merit yet. Most arguments will begin with a claim. (*E.g., You should buy an aluminum can crusher for your home.*)

2. *Data*   These are the facts that one uses as a foundation in order to later establish validity to the claim. (*E.g., Environmentally, recycling aluminum cans produces 95% less air pollution.*)

3. *Warrant*   This is the link made from the data to the claim in order to authorize or validate it. (*E.g., Buying an aluminum can crusher for your home will be better for the environment by causing less air pollution.*)

Optional Elements of An Argument:

4. *Backing*   These are facts that can be used to give credit to the warrant. This can be done by providing evidence to the statement made or by making another statement that adds credibility to the warrant. (*E.g., You told me last week that you wanted to be more responsible to the environment after reading about the new environmental legislation.*)

5. *Rebuttal*   These are statements that recognize restrictions or limitations to the claim. (*E.g., Unless of course you have changed your mind about being more responsible to the environment*)

6. *Qualifier*   These are words that qualify how certain you are about your claim. Words such as 'certain', 'probable', and 'presumably', etc. are typical examples. (*E.g., I am certain that the can crusher will be more friendly to the environment, which is what I presume you want.*)

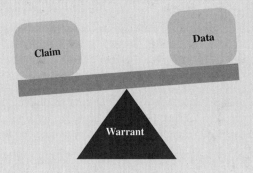

Discussion Discuss the following statements or claims in class. Some are *true* statements; some are *false*, while others may be *either*. What is your opinion and why? Try to remember Toulmin's model when putting your argument forward.

- Ethics is not the same as feelings.
- Ethics is following the law.
- Ethics is not a science.
- Ethics is following culturally accepted norms.
- In some cases, it may be ethical to make and act on an unethical decision.
- Designing and/or selling a gun is unethical.

Ethical and Moral Case

Discuss the following scenario that is divided into two parts. Follow the steps described previously for approaching ethical dilemmas, moral reasoning, and presenting an argument. Refer to the NSPE Code of Ethics as well as the appropriate moral frameworks. Justify your reasoning and be prepared to debate your viewpoint to those that disagree with you in the class!

1. High Concept Manufacturing (HCM) has an engineering plant in a small town that employs 12.4% of the community. It provides approximately $10 million dollars of salaries to its community workers and pays $2 million in taxes to the local government. As a consequence of some of its manufacturing procedures, the HCM plant releases bad smelling fumes. These fumes annoy HCM's residential neighbors, hurt the local tourism trade, and have been linked (although not conclusively) to a rise in asthma in the area. The financial impact of this to the town is estimated to be around $3 million as a result of a decrease in tourism and lower house prices. The town is considering issuing an ultimatum (final warning) to HCM; "Clean up your plant, or we will fine you $1 million." HCM had previously made it known that the business will close down and go somewhere else if it is fined by the town. What should the town do?

2. There will a town meeting where all concerned parties have agreed to attend and discuss the matters given. You are an engineer and a respected member of the town who has been recently offered an excellent job opportunity at HCM. You have signed a contract with HCM, and you are officially one of their new employees. However, this is as of yet not public knowledge. HCM asks you to try to convince the town to drop the case and that the town is better off with HCM's presence. What are you going to do?

References

The following articles/websites were used in preparing this lab:
1. Toulmin, S., *The Uses of Argument*. Cambridge University Press, 1958.
2. http://www.scu.edu/ethics/practicing/decision/framework.html
3. http://www.ethicsandbusiness.org/pdf/strategy.pdf
4. http://ocw.usu.edu/English/intermediate-writing/english-2010/2010/toulmins-sohema ∎

1.7 PROBLEMS

1.7.1 Team Activities

1. List the steps in the design process.
2. Define each step in the design process.
3. What is the difference between customer statement and problem definition?
4. What is the difference between the specification step and defining the problem step in the design process?
5. What is function analysis and how is it different from the definition of the problem?
6. List three factors that market analysis achieves.
7. Why does function analysis precede the conceptualization step?
8. Give examples of a mock-up.
9. Why is scheduling important?

1.7.2 Individual Activities

1. Figure 1.10 shows the percent of cost committed and incurred as a function of time during the design of a product. The committed cost is the amount of money allocated for the manufacturing of the product, while the incurred cost is the amount of money spent on the design.
   a. Explain your findings from the figure.
   b. Discuss why it is more important to spend time and money on the early stages of design than on the late stages of design.

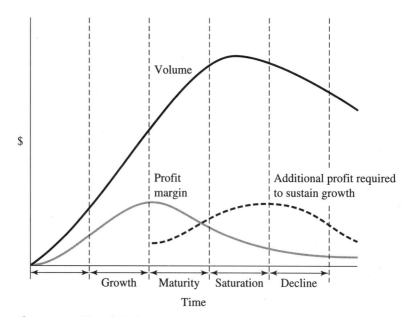

Figure 1.10   Manufacturing cost.

2. Describe in detail the various types of design.

3. List several common sources of engineering failures.

4. Explain the difference between engineering design and the engineering design process.

5. Identify several commonalities and differences among Figures 1.1 through 1.4.

6. List four factors that may be used to determine quality, and discuss the following statement in light of your listing: "Quality cannot be built into a product unless it is designed into it."

7. Consider the following two statements:
   - What size SAE grade 5 bolt should be used to fasten together two pieces of 1045 sheet steel, each 4 mm thick and 6 cm wide, which are lapped over each other and loaded with 100 N?
   - Design a joint to fasten together two pieces of 1045 sheet steel, each 4 mm thick and 6 cm wide, which are lapped over each other and loaded with 100 N.
     (a) Explain the difference between the two statements.
     (b) Change a problem from one of your engineering science/physics classes into a design problem.

8. Explain the engineering design process to a high school student.

9. Define professionalism.

10. Read through the ASME code of ethics and list three findings.

11. The following ethical scenario was obtained from http://www.cwru.edu/wwwethics.
    You are an engineer charged with performing safety testing and obtaining appropriate regulatory agency or outside testing laboratory approvals of your company's product. The Gee Whiz Mark2 (GWM2) has been tested and found compliant with both voluntary and mandatory safety standards in North America and Europe. Because of a purchase-order error and subsequent oversights in manufacture, 25,000 units of GWM2 were built that are not compliant with any of the North American or European safety standards. A user would be more vulnerable to electric shock than from a compliant unit. Under some plausible combinations of events, the user of the bad unit could be electrocuted. Retrofitting these products to make them compliant is not feasible because the rework costs would exceed the profit margin by far. The company agrees that because of this defect the agency safety labels will not be attached to the bad units, as per the requirements of the several agencies. Only two options exist: (a) Scrap the units and take the loss, or (b) sell the units. An employee of the company notes that many countries have no safety standards of any kind for this type of product. It is suggested that the bad units be marketed in these countries. It is pointed out that many of these nations have no electrical wiring code; or if codes exist they are not enforced. The argument is thus advanced that the bad GWM2 units are no worse than the modus operandi of the electrical practice of these countries and their cultural values. Assume that no treaties or export regulations would be violated.
    a. What would be your recommendation?
    b. Suppose one of the countries under consideration was the country of origin for you or your recent ancestors. Would this affect your recommendations?

    c. Now suppose you are not asked for a recommendation, only an opinion. What is your response?

    d. Suppose it is suggested that the bad units be sold to a third party, who would very likely sell the units to these countries. Your comments?

    e. You are offered gratis one of the bad units for your use at home, provided that you sign a release indicating your awareness of the condition of the unit and that it is given to you as a test unit. Assume you cannot retrofit it and that the product could be very useful to you. Would you accept the offer?

    f. Suppose it is suggested that the offer stated in part (e) be made to all employees of the company. Your comments?

1.8 Selected Bibliography

AMBROSE, S. A. and AMON, C. H. "Systematic Design of a First-Year Mechanical Engineering Course at Carnegie Mellon University." *Journal of Engineering Education*, pp. 173–181, 1997.

BUCCIARELLI, L. L. *Designing Engineers*. Cambridge, MA: MIT Press, 1996.

BURGER, C. P. "Excellence in Product Development through Innovative Engineering Design." *Engineering Productivity and Valve Technology*, pp. 1–4, 1995.

BURGHARDT, M. D. *Introduction to Engineering Design and Problem Solving*. New York: McGraw-Hill, 1999.

CROSS, N. *Engineering Design Methods: Strategies for Product Design*. New York: Wiley, 1994.

CROSS, N., CHRISTIAN, H., and DORST, K. *Analysing Design Activity*. New York: Wiley, 1996.

DHILLON, B. S. *Engineering Design: A Modern Approach*. Toronto: Irwin, 1995.

DIETER, G. *Engineering Design*. New York: McGraw-Hill, 1983.

DYM, C. L. *Engineering Design: A Synthesis of Views*. Cambridge, UK: Cambridge University Press, 1994.

EEKELS, J., and ROOZNBURG, N. F. M. "A Methodological Comparison of Structures of Scientific Research and Engineering Design: Their Similarities and Differences." *Design Studies*, Vol. 12, No. 4, pp. 197–203, 1991.

FLEDDERMANN, C. B. *Engineering Ethics*. Upper Saddle River, NJ: Prentice Hall, 1999.

HENSEL, E. "A Multi-Faceted Design Process for Multi-Disciplinary Capstone Design Projects." *Proceedings of the 2001 American Society for Engineering Education Annual Conference and Exposition*, Albuquerque, NM, 2001.

HILL, P. H. *The Science of Engineering Design*. New York: McGraw-Hill, 1983.

HORENSTEIN, M. N. *Design Concepts for Engineers*. Upper Saddle River, NJ: Prentice Hall, 1999.

JOHNSON, R. C. *Mechanical Design Synthesis*. Huntington, NY: Krieger, 1978.

WATSON, S. R. "Civil Engineering History Gives Valuable Lessons." *Civil Engineering*, pp. 48–51, 1975.

JANSSON, D. G., CONDOOR, S. S., and BROCK, H. R. "Cognition in Design: Viewing the Hidden Side of the Design Process." *Environment and Planning B, Planning and Design*, Vol. 19, pp. 257–271, 1993.

KARUPPOOR, S. S., BURGER, C. P., and CHONA, R. "A Way of Doing Design." *Proceedings of the 2001 American Society for Engineering Education Annual Conference and Exposition*, Albuquerque, NM, 2001.

KELLEY D. S., NEWCOMER, J. L., and MCKELL, E. K. "The Design Process, Ideation and Computer-Aided Design." *Proceedings of the 2001 American Society for Engineering Education Annual Conference and Exposition*, Albuquerque, NM, 2001.

PAHL, G., and BEITZ, W. *Engineering Design: A Systematic Approach*. New York: Springer-Verlag, 1996.

PUGH, S. *Total Design*. Reading, MA: Addison-Wesley, 1990.

RAY, M. S. *Elements of Engineering Design*. Englewood Cliffs, NJ: Prentice Hall, 1985.

RADCLIFFE, D. F., and LEE, T. Y. "Design Methods used by Undergraduate Students," *Design Studies*, Vol. 10, No. 4, pp. 199–207, 1989.

Ross, S. S. *Construction Disasters: Design Failures, Causes and Preventions*. New York: McGraw-Hill, 1984, pp. 303–329.

Sickafus, E. N. *Unified Structured Inventive Thinking*. New York: Ntelleck, 1997.

Siddall, J. N. "Mechanical Design." *ASME Transactions: Journal of Mechanical Design*, Vol. 101, pp. 674–681, 1979.

Suh, N. P. *The Principles of Design*. Oxford, UK: Oxford University Press, 1990.

Ulman, D. G. *The Mechanical Design Process*. New York: McGraw-Hill, 1992.

Ulrich, K.T. and Eppinger, S. D. *Product Design and Development*. New York: McGraw-Hill, 1995.

Vidosic, J. P. *Elements of Engineering Design*. New York: The Ronald Press Co., 1969.

Walton, J. *Engineering Design: From Art to Practice*. New York: West Publishing Company, 1991.

Whitebeck, C. http://web.mit.edu/ethics/www/essays/probcase.html

CHAPTER · 2

Essential Transferable Skills

Design is a difficult challenge as it requires the designers to predict the success of their eventual products at all levels and ensure that they are used as intended. This requires the designers to possess many essential transferable skills in addition to their technical and creative expertise. (RAGMA IMAGES/Shutterstock)

2.1 OBJECTIVES

By the end of this chapter, you should be able to

1. Identify the essential skills that are prerequisite to the design process.
2. Appreciate the importance and dynamics of working within teams.
3. Develop a project schedule utilizing existing tools.
4. Practice and improve your research and communication skills.

Before discussing the design process, it is important to state some necessary prerequisites. Transferable skills are used throughout the entire design process and indeed are needed throughout the entire life of a professional engineer no matter what specialization or specific engineering career path they choose. This chapter will focus on the following skills.

Working in Teams—The dynamics of a team and how to work within one is vital to the success of any project.

Scheduling—Necessary to manage a project and arrive at a successful conclusion in a time efficient manner. It also focuses members of a team to diverge and converge at relevant stages of the project in order to provide an integrated solution.

Research Skills—Important for collecting data, gathering information, and keeping up with the latest both technically and competitively.

Technical Writing—Communication and reporting is a vital component of any design and is sometimes overlooked as a minor burden that needs to be done. However, even with the most innovative design that may revolutionize the way people live, if it is not reported to the world, it will remain an unknown entity never to be adopted or adapted and ultimately will become a failure.

Presentation Skills—Just as important as technical writing and sometimes more powerful. This skill has provides the ability to reach one important person who will agree to support an idea all the way to the mass population who will buy it. As such, it is an extremely powerful reporting and marketing tool. As the presenter is directly communicating with potentially interested parties, he/she has the distinct advantage and opportunity to convince and promote an idea which may have otherwise been overlooked.

2.2 WORKING IN TEAMS

Working in teams is an inevitable necessity. In a world where time to market has become a competitive component, a team of individuals will undoubtedly complete a project in a much shorter space of time than a single person would take. Furthermore, in many cases, the breadth of expertise and knowledge that is required to visualize and realize a project makes it impossible for a single person to achieve even if time was not an issue.

Mohrman and Mohrman define a team as a collection of individuals whose work is interdependent and who are collectively responsible for accomplishing a performance outcome.[1] Thus, a collection of individuals in a music consort doesn't necessarily constitute a team, although all of the individuals are in the same place at the same time for the same purpose. The keyword is *collectively*. A group of students assembled at the start of the semester doesn't compose a team unless the students are working collectively toward accomplishing an outcome.

A team can be likened to a meal. Both need 'ingredients' and a 'recipe' to be successful. The 'ingredients' are the individual team members, whereas the 'recipe' is the dynamics of the team: how they interact with each other and how they behave. Without both of these items in place, a team, like a meal, will not have a satisfactory outcome.

2.2.1 Forming a Team

When picking appropriate individuals to form a successful team, it is important to select individuals that complement each other. Naturally, each person thinks and behaves in preferred ways that are unique to that individual. These dominant thinking styles are the results of the native personality interacting with family, education, work, and social environments. People's approaches to problem solving, creativity, and communicating with others are characterized by their thinking preferences. For example, one person may carefully analyze a situation before making a rational, logical decision based on the available data. Another may see the same situation in a broader context and look for several alternatives. One person will use a very detailed, cautious, step-by-step procedure. Another has a need to talk the problem over with people and will solve the problem intuitively.

Several models have been proposed on how the human brain works. One of the well-known models is the Herrmann model. Herrmann developed a metaphorical model of the brain that consists of four quadrants. Although all of us are using all four quadrants, some individuals may have more use of certain quadrants of the brain. The following is a listing of these quadrants and their characteristics:

Upper left: The characteristics of this quadrant are analytical, logical, quantitative, and fact based.

Lower left: The characteristics of this quadrant are organized, planned, detailed, and sequential.

Upper right: The characteristics of this quadrant are holistic, intuitive, synthesizing, and integrating.

Lower right: The characteristics of this quadrant are emotional, social, and communicative.

Lab 3 will help students to understand these four thinking styles in more detail as well as allow them to better understand their own thought processes. Forming a team with all four thinking styles will compose a better team, and this lab should be used for this purpose before the design teams are created within the classroom.

---

[1]Mohrman, S. A., and Mohrman, A. M. *Designing and Leading Team-Based Organization: A Workbook for Organizational Self Design.* San Francisco: Jossey-Bass, 1997.

LAB 3: Ice Breaking—Forming Teams

This lab introduces these models and provides a tool with which students can better understand their thought processes. When students understand their own thought processes and those of others, they work better in a group.

The objectives of this lab are as follows:

1. Introduce the model of the brain developed by Ned Herrmann.
2. Present a preliminary self-test to quantify the strength of the four quadrants.
3. Introduce team formation. After the students test which thinking preference they may have, it becomes a tool when the teams form to make a full brain (i.e., one student from A, another from B, and so on).

Each person thinks and behaves in preferred ways that are unique to that individual. These dominant thinking styles are the results of the native personality interacting with family, education, work, and social environments. People's approaches to problem solving, creativity, and communicating with

In this lab, we will discuss such team-building skills as brainstorming. Each person thinks and behaves in preferred ways that are unique to that individual. Harnessing each individual's thinking style can lead to productive brainstorming sessions. (Carlosseller/Shutterstock)

others are characterized by their thinking preferences. For example, one person may carefully analyze a situation before making a rational, logical decision based on the available data. Another may see the same situation in a broader context and look for several alternatives. One person will use a very detailed, cautious, step-by-step procedure.

Another has a need to talk the problem over with people and will solve the problem intuitively. Ned Herrmann's metaphorical model divides the brain into left and right halves and into cerebral and limbic hemispheres, resulting in four distinct quadrants. The Hermann profile results have a neutral value. There are no right or wrong answers.

The questionnaire presented in Table L3.1 will help you investigate your own thinking style. Questionnaires like this one are available from specialized companies such Ned Herrmann's for a nominal charge per student. The idea here is to have this activity and questionnaire as a preliminary step for people who have more interest in finding more and sufficient for people who just want some idea.

The questions in this questionnaire are based on data available in Lumsdaine's book and material distributed through the NSF (National Science Foundation) workshop on introductory engineering design at the Central Michigan University, 1999, presented by Frank Maraviglia.[2]

Procedure

In the rating box to the right, after Questions 1–13, write the numbers 4, 3, 2, or 1, where a 4 indicates most likely and 1 indicates least likely. Use each number ONLY ONCE for each question. There is no right or wrong answer; answer to the best of your knowledge and experience.

To find out the totals you need to

1. Multiply the total number in 14 by 4 and then add the results to the I bracket. Once you have the total sum, call it A.
2. Multiply the total number in 15 by 4 and then add the results to the II bracket. Once you have the total sum, call it B.
3. Multiply the total number in 16 by 4 and then add the results to the III bracket. Once you have the total sum, call it C.
4. Multiply the total number in 17 by 4 and then add the results to the IV bracket. Once you have the total sum, call it D.

In items 14 to 17 you will find four lists. Check the number of items in each list that you enjoy doing. Add up the number of checked items and write the number in the space provided for each item.

---

[2]Based on Lumsdaine, E., and Lumsdaine, M., *Creative Problem Solving*. New York: McGraw-Hill, 1995 and NSF Sponsored Workshop on Introductory Engineering Design, J. Nee (editor). Central Michigan University, 1999.

TABLE L3.1 Questionnaire

	Question	Options	Rating
e.g.	This is an example question	a. Provide a Rating of 1, 2, 3, OR 4 in each box on the right b. 4 indicates most likely and 1 indicates least likely c. Use each number only ONCE for each question d. There is no right or wrong answer. Answer to the best of your knowledge	3 4 1 2
1	In a project setup you prefer to perform the following function:	a. Looking for data and information b. Developing a systematic solution and directions c. Listening to others and sharing ideas and intuitions d. Looking for the big picture and context, not the details	
2	While solving your homework you	a. Organize the information logically in a framework but not down to the least detail b. Do detailed homework problems neatly and conscientiously c. Motivate yourself by asking why and by looking at personal meaning d. Take the initiative in getting actively involved to make learning more interesting	
3	In studying a new material you prefer	a. Listening to the informational lecture b. Testing theories and procedures to find flaws and shortcomings c. Reading the preface of the book to get clues on the purpose d. Doing simulations and ask what-if questions	
4	Do you like	a. Reading textbooks b. Doing lab work step-by-step c. Talking about, seeing, testing, and listening d. Using visual aids instead of words	
5	In studying new material you like to	a. Analyze example problems and solutions b. Write a sequential report on the results of a lab experiment c. Do hands-on learning by touching and seeing d. Take an open-ended approach and find several solutions	
6	When and after solving a problem you	a. Think through ideas rationally b. Find practical uses of knowledge learned; theory is not enough c. Use group study opportunities and group discussion d. Appreciate the beauty in the problem and the elegance of the solution	
7	Do you prefer to	a. Do research using the scientific method b. Use computers with tutorial software c. Use group study opportunities and group discussion d. Lead brainstorming sessions in which wild ideas are accepted	
8	In project execution, which of the following do you prefer to do?	a. Make up a hypothesis and then test it to find out if it's true b. Plan, schedule, and execute projects according to a set time c. Keep a journal to record what you have seen in the experiment d. Experiment and play with ideas and possibilities	
9	Generally speaking, you absorb new ideas best by	a. Applying them to concrete situations b. Concentrating and conducting careful analysis c. Contrasting them with other ideas d. Relating them to current or future activities	
10	When you read self-help articles you pay most attention to	a. The ideas that are drawn from the information b. The truth of the finding backed up by information c. Whether or not the recommendations can be accomplished d. The relation of the conclusion to your experience	
11	When you hear people arguing you favor the side that	a. Presents ideas based on facts and logic b. Expresses the argument most forcefully and concisely c. Reflects your personal opinion d. Projects the future and shows the total picture	
12	When you make a new choice you rely most on	a. Reality and the present rather than future possibilities b. Detailed and comprehensive studies c. Talking to people d. Intuition	
13	When you approach a problem you are likely to	a. Try to relate to a broader problem or theory b. Try to find the best procedure for solving it c. Try to imagine how others might solve it d. Look for ways to solve the problem quickly	
	Add up the columns vertically in the space provided (not including example question)		I | II | III | IV

TABLE L3.1 Questionnaire

14. Total: [   ]
    — Looking for data and information; doing library searches
    — Organizing information logically in a framework
    — Listening to informational lectures
    — Reading textbooks
    — Studying sample problems
    — Thinking through ideas
    — Making up a hypothesis and then testing it to find out if it is true
    — Judging ideas based on facts, criteria, and logical reasoning
    — Doing technical and financial case studies
    — Knowing how much things cost
    — Using computers for math and information processing

15. Total: [   ]
    — Following directions carefully, instead of improvising
    — Testing theories and procedures to find flaws and shortcomings
    — Doing lab work step by step
    — Listening to detailed lectures
    — Taking detailed comprehensive notes
    — Studying according to a fixed schedule in an orderly environment
    — Making up a detailed budget
    — Practicing new skills through frequent repetition
    — Writing a how-to manual about a project
    — Using computers with tutorial software
    — Finding practical uses of knowledge learned

16. Total: [   ]
    — Listening to others and sharing ideas
    — Keeping a journal to record feelings and spiritual values
    — Studying with background music
    — Respecting others' rights and views; people are important, not things
    — Using visual clues to make use of body language
    — Doing hands-on learning by touching and using a tool or object
    — Watching drama movies over adventure movies
    — Learning through sensory input (i.e., moving, feeling, smelling, testing, listening)
    — Motivating yourself by asking why and by looking for personal meaning
    — Traveling to meet other people and learn about their culture

17. Total: [   ]
    — Looking for the big picture and context, not details, of a new topic
    — Taking the initiative in getting actively involved to make learning more interesting
    — Doing open-ended problems and finding several possible solutions
    — Thinking about trends
    — Thinking about the future and making up long-range goals
    — Admiring the elegance of inventions, not the details
    — Synthesizing ideas and information to come up with something new
    — Relying on intuition to find solutions
    — Predicting what future technology may look like
    — Looking for alternative ways of accomplishing a solution
    — Using pictures instead of words when learning

To find out the totals you need to

1. Multiply the total number in 14 by 4 and then add the results to the I bracket. Once you have the total sum, call it A.
2. Multiply the total number in 15 by 4 and then add the results to the II bracket. Once you have the total sum, call it B.
3. Multiply the total number in 16 by 4 and then add the results to the III bracket. Once you have the total sum, call it C.
4. Multiply the total number in 17 by 4 and then add the results to the IV bracket. Once you have the total sum, call it D.

Sample Result

The following is a sample of one student's questionnaire results:

```
[1]  [3]  [4]  [2]
[3]  [4]  [1]  [2]
[4]  [3]  [1]  [2]
[2]  [1]  [4]  [3]
[4]  [1]  [3]  [2]
[4]  [2]  [3]  [1]
[4]  [2]  [3]  [1]
[3]  [4]  [1]  [2]
[3]  [4]  [2]  [1]
[4]  [1]  [3]  [2]
[3]  [2]  [4]  [1]
[4]  [1]  [3]  [2]
[3]  [4]  [1]  [2]
 I    II   III   IV
```

- Total [42] [32] [33] [23]

- 14 Total [7]

- 15 Total [5]

- 16 Total [4]

- 17 Total [4]

The sum is expressed as

$A = (4 \times 7) + 42 = 70$
$B = (4 \times 5) + 32 = 52$
$C = (4 \times 4) + 33 = 49$
$D = (4 \times 4) + 23 = 39$

To express the finding graphically, use an Excel spreadsheet. Then use the radar chart within Excel to express the values graphically, as shown in Figure L3.1. People with an A preference are analytical, rational, technical, logical, factual, and quantitative. B-preference people are procedural, scheduled, conservative, organized, sequential, reliable, tactical, and administrative. C-preference students are supportive, interpersonal, expressive, sensitive, symbolic, musical, and reaching out. D-preference thinkers are more strategic, visual, imaginative, conceptual, and simultaneous.

Discussion

Apparently, the student shown in the example has a strong preference for the A quadrant—analytical. Studies have shown that engineering faculty are predominantly A thinkers. Students change by virtue of practice to A thinkers by the end of their careers. Design students are generally D thinkers.

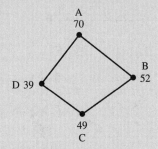

Figure L3.1   Sketch of student's thinking preference.

Everyone is creative and can learn to be more creative. Everyone can learn to use the D-quadrant thinking abilities of the brain more frequently and more effectively.

So don't assume that creativity can't be learned and only a few have that ability. Another false assumption is the perception that an intelligent mind is a good thinker. According to Edward de Bono, highly intelligent people who are not properly trained may be poor thinkers for a number of reasons.

1. They can construct a rational, well-argued case for any point of view and thus do not see the need to explore alternatives.
2. Because verbal fluency is often mistaken for good thinking, they learn to substitute one for the other.
3. Their mental quickness leads them to jump to conclusions from only a few data points.
4. They mistake understanding with quick thinking and slowness with being dull witted.
5. The critical use of intelligence is usually more satisfying than the constructive use. Proving someone wrong gives instant superiority but doesn't lead to creative thinking.

Lab 3 Problems

1. Which quadrant would you consider a dominant quadrant for the following well-known figures?
   (a) Newton
   (b) Galileo
   (c) Einstein
   (d) Leonardo da Vinci
   (e) Malcolm X
   (f) Bach
   (g) Martin Luther King, Jr.
   (h) George Bush, Jr.
   (i) Bill Clinton
   (j) Ben Franklin
   (k) Gandhi

2. Would you be able to change your thinking style? Why or why not?
3. If you are organizing a brainstorm session, which thinking styles would you like to have in the session and why?
4. Which role would you like to play in a team? Justify your answer based on your findings in the questionnaire. ∎

2.2.2 Dynamics of a Team

Once a team has been formed, the dynamics of the team contributes to it success or failure. Regardless of the performance outcome, McGoutry and De Muse observed that all teams are characterized by the following features.

1. A dynamic exchange of information and resources among team members.
2. Task activities coordinated among individuals in the group.
3. A high level of interdependence among team members.
4. Ongoing adjustments to both the team and individual task demands.
5. A shared authority and mutual accountability for performance.

Many students dread group projects. Past experience with poorly functional teams and a lack of school system training may have built this fear in students.

Larson and LaFasto studied the characteristics of effective functional teams. They reported that successful teams have the following characteristics.

1. A clear, challenging goal; this goal gives the group members something to shoot for. The goal is understood and accepted by the entire group.
2. A result-driven structure; the roles of each member are clear, a set of accountability measure is defined, and an effective communication system is established.
3. Competent and talented team members.
4. Commitment; team members put the team goals ahead of individual needs.
5. Positive team culture. This factor consists of four elements:
   (a) Honesty
   (b) Openness
   (c) Respect
   (d) Consistency in performance
6. Standard of excellence.
7. External support and recognition. Effective teams receive the necessary resources and encouragement from outside the group.
8. Effective leadership.

Design can be considered as a social activity in which a collective effort from a team is put together to produce the required output. Keep in mind that successful teams do not occur automatically or overnight. Effort and time need to be devoted to nurturing a successful team. Also, there are stages according to which teams evolve:

Forming: At this stage, the group members still work as individuals; they do not contribute to the group as a whole but look out for themselves.

Storming: At this stage, the group realizes that the task requires collective contribution and not much has been done. This prompts disagreements, blaming, and impatience with the process. Some members try to do it all on their own and avoid collaboration with team members.

Norming: When the team's objectives are worked out collectively, the common problems or goals begin to draw individuals together into a group, although the sense of individual responsibility is still very strong. Conversation among the team members helps direct efforts toward better teamwork and increase members' sense of responsibility for the team objective.

Performing: At this stage, the team members have accepted each others' strengths and weakness and have defined workable team roles.

McGourty, et al. and DeMeuse and Erffmeyer present four behaviors that are needed for effective team performance.

1. *Communication team behavior:* Team members need to create an environment in which all members feel free to speak and listen attentively. The following behavioral practices for both roles must be followed:
   a. Listen attentively to others without interpreting.
   b. Convey interest in what others are saying.
   c. Provide others with constructive feedback.
   d. Restate what has been said to show understanding.
   e. Clarify what others have said to ensure understanding.
   f. Articulate ideas clearly and concisely.
   g. Use facts to get points across to others.
   h. Persuade others to adopt a particular point of view.
   i. Give compelling reasons for ideas.
   j. Win support from others.
2. *Decision-making team behavior:* Decision making is done by the team, not for the team. The following behavioral practices must be followed.
   a. Analyze problems from different points of view.
   b. Anticipate problems and develop contingency plans.
   c. Recognize the interrelationships among problems and issues.
   d. Review solutions from opposing perspectives.
   e. Apply logic in solving problems.
   f. Play a devil's advocate role when needed.

g. Challenge the way things are being done.
h. Solicit new ideas from others.
i. Accept change.
j. Discourage others from rushing to conclusions without facts.
k. Organize information into meaningful categories.
l. Bring information from outside sources to help make decisions.

3. *Collaboration team behavior:* Collaboration is the essence of team work; it involves working with others in a positive, cooperative, and constructive manner. The following behavioral practices must be followed.
    a. Acknowledge issues that the team needs to confront and resolve.
    b. Encourage ideas and opinions even when they are different from your own.
    c. Work toward solutions and compromises that are acceptable to all.
    d. Help reconcile differences of opinion.
    e. Accept criticism openly and nondefensively.
    f. Share credit for success with others.
    g. Cooperate with others.
    h. Share information with others.
    i. Reinforce the contribution of others.

4. *Self-management team behavior:* The following behavioral practices must be followed.
    a. Monitor progress to ensure that goals are met.
    b. Create action plans and timetables for work session goals.
    c. Define task priorities for work sessions.
    d. Stay focused on the task during meetings.
    e. Use meeting time efficiently.
    f. Review progress throughout work sessions.
    g. Clarify roles and responsibilities of others.

Each of these behaviors when followed will assure an effective team. It is important that distinct team roles be defined. The following roles apply to a team of four members:

1. *Captain*, who possesses behaviors and skills described in self-management team behavior.
2. *Chief engineer*, who possesses behaviors in decision-making team behavior.
3. *Human resources person*, who possesses behaviors described in collaboration team behavior.
4. *Spokesperson*, who possess behaviors described in communication team behavior.

Lab 4 covers the practical side of the dynamics of a team. Once students have completed Lab 3 and formed their teams based on the four thinking styles, they should then continue onto Lab 4 in order to provide them with an induction to working together as a unit.

LAB 4: Team Dynamics

Purpose

Design (as well as other major tasks) requires individuals to work in teams. Major companies (as well as small businesses) recognize the need for employees to work in teams to achieve a goal. Managerial initiatives, such as total quality management (TQM) and process reengineering, recommend teams as the preferred way to organize and accomplish work. Concurrent engineering relies on cross-functional teams to enhance product development and innovation. Many undergraduate classes feature team assignments to help students develop the skills required for a successful career. However, especially in engineering, little class time is devoted to solving issues as a team. Thus the objectives of this lab are as follows:

1. Discuss the dynamics of team formation.
2. Present methodology for effective teams.

Theory

As we discussed previously in Section 2.2, a team is a collection of individuals whose work is interdependent and who are collectively responsible for accomplishing a performance outcome. Thus, a

(Dmitriy Shironosov/Shutterstock)

collection of individuals in a music consort doesn't constitute a team, although all the individuals are in the same place at the same time for the same purpose. The keyword is *collectively*. A group of students assembled at the start of the semester doesn't compose a team unless the students are working collectively toward accomplishing an outcome.

This lab should be used in conjunction with Section 2.2.2 on dynamics of a team. If you have not done so already, then refer to the classification of teams by McGoutry and De Meuse in that section before continuing

Although this lab addresses some important points in team dynamics, there are many topics that need to be discussed. Some of these topics are addressed as part of the critical thinking activities. It is recommended that once students finish the exercise in the procedure section they devote time and effort to addressing the questions in the critical thinking section.

Procedure

1. Arrange in teams of four members each. Make sure that all thinking styles are represented in the team as much as possible.

2. Map each of the team behaviors presented in Section 2.2 to a thinking style. Discuss your findings.

3. Once you have done the mapping, assign roles to each team member based on the recommendations in the previous section. Ask the human resources person to play the role of reflector, who will write a report describing the team dynamics during the brainstorming session. Assign the spokesperson to be a recorder of the actions and ideas generated during the brainstorming session. The chief engineer, once the session is over, will judge all the ideas and submit a report to the captain (CEO) of the team. The captain, once the exercise is over, will submit a report detailing the team activity to the board of directors (the classmates and instructor).

4. Consider the following for a brainstorming exercise:
   The energy crisis in California in the summer of 2001 has decision makers in your community concerned about a similar crisis. Brainstorm the following energy conservation topics:
   a. How can houses be built or improved to be more energy efficient?
   b. How can your local community transportation system be changed to be more energy efficient? ∎

2.3   SCHEDULING

Development of a new product according to the design process is always limited by the time available for the entire process. Over the years, several procedures have been proposed to manage and plan for an artifact. Many projects have failed due to lack of detail. Historically, engineering projects were typically created by one engineer, working alone. The designer, the drafter, and the planner were the same person. In recent years, however, advances in technology require that teams of engineers and technical helpers work beside each other to accomplish a project. In the following subsections, we discuss existing techniques for scheduling.

2.3.1 Gantt Chart

The Gantt chart was introduced by Henry L. Gantt and Frederick Taylor in the early 1900s to facilitate project management. The Gantt chart is in the form of a bar chart and is established as follows:

1. List all events or milestones of the project in an ordered list.
2. Estimate the time required to establish each event.
3. List the starting time and end time for each event.
4. Represent the information in a bar chart.

An example of the Gantt chart is shown in Figure 2.1. Each event is allocated a start time and duration shown by the shaded rectangles. If certain events can begin in parallel, they are given the same start date as can be seen with the Market Analysis and Specifications. However, certain events cannot begin until prior events have been completed. An example of this in the figure is Prototype testing which begins in month 8 but obviously only after the prototype construction event has been completed at the end of month 7. Gantt charts can organize a project to arrive at its successful conclusion in an efficient manner by identifying which events are independent of each other and starting them at their earliest possible time. Commercial software is available that can help in developing a Gantt chart for the project (e.g., Microsoft Project Manager™ or Oracle Primavera™). Lab 5 introduces the creation of Gantt charts in Microsoft Project Manager. Even if students do not have this software, they should attempt the lab either using an alternative software package or manually, as the case study included provides good training and practice in creating the charts.

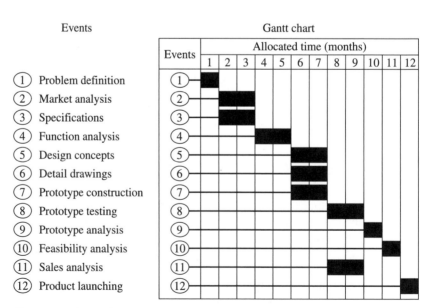

Figure 2.1   Gantt chart.

2.3.2 CPM/PERT

The critical path method (CPM) and the program evaluation and review technique (PERT), which were developed during the 1950s and 1960s, are the two most widely used approaches for scheduling projects. PERT was developed under the guidance of the U.S. Navy Special Projects Office by a team that included members from the Lockheed Missile Systems Division and the consulting firm Booz, Allen, and Hamilton. The technique was developed to monitor the efforts of 250 main contractors and 9000 subcontractors who were concerned with the Polaris missile project. CPM was the result of the efforts of the Integrated Engineering Control Group of E.I. duPont de Nemours & Company, and it was developed to monitor activities related to design and construction. A survey of manufacturing companies in the United States revealed that CPM/PERT is used over 65% of the time, among all other methods. A CPM/PERT project generally has the following characteristics:

1. There are clearly defined activities or jobs whose accomplishment results in project completion.
2. Once started, the activity or job continues without interruption.
3. The activities or tasks are independent, which means they may be started, stopped, and performed individually in a prescribed sequence.
4. The activities or jobs are ordered and they follow each other in a specified manner.

Even though the CPM and PERT techniques were developed independently, the basic theory and symbology in both techniques are essentially the same.

2.3.3 CPM/PERT Definitions

Several symbols, terms, and definitions are used in developing the CPM/PERT networks:

1. *Event (node):* This represents a point in time in the life of a project. An event can be the beginning or the end of an activity. A circle is used to represent an event. Generally, each network event is identified with a number.
2. *Activity:* This is an effort needed to carry out a certain portion of the project.
3. *Network paths:* These are the paths used (or needed) for reaching the project termination point, or the event from the project starting point or event.
4. *Critical path:* This is the longest path with respect to time duration through the PERT/CPM network. In other words, the critical path creates the largest sum of activity duration of all individual network paths.
5. *Earliest event time (EET):* This is the earliest time at which an event occurs, providing all proceeding activities are accomplished within their estimated times.
6. *Latest event time (LET):* This is the latest time at which an event could be reached without delaying the predicted project completion date.
7. *Total float:* This is the latest time of an event, minus the earliest time of the preceding event, and the duration time of the in-between activity.

Figure 2.2 illustrates an example of a CPM or PERT chart. The longest path or the critical path is Task A-B-E-H which takes 14 days. As task D would finish after 12 days (three days for Task A followed by the allocated nine days for Task D), there would be two days remaining until the next event, which in this case is the end of the project. Therefore there is a two-day float for Task D. Similarly, it can be ascertained that there is a one-day float to complete tasks C, F, and G before Task H can begin.

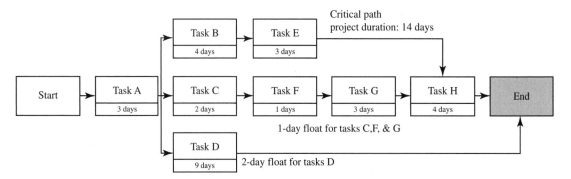

Figure 2.2   CPM/PERT chart.

2.3.4 CPM/PERT Network Development

Steps involved in constructing a CPM network are

1. Break down the design into individual activities and identify each activity.
2. Estimate the time required for each activity.
3. Determine the activity sequence.
4. Construct the CPM network, using the defined symbols.
5. Determine the critical path of the network.

Similarly, the steps involved in developing a PERT network are

1. Break down the design into individual activities and identify each activity.
2. Determine the activity sequence.
3. Construct the PERT network, using the defined symbols.
4. Obtain the expected time to perform each activity, using the following weighted average formula:

$$T_e = \frac{x + 4y + z}{6}$$

where

Te = the expected time for the activity

x = the optimistic time estimate for the activity

y = the most likely time estimate for the activity

z = the pessimistic time estimate for the activity

5. Determine the network critical path.
6. Compute the variance associated with the estimated expected time of Te of each activity using the formula:

$$S^2 = \left(\frac{z - x}{6} \right)^2$$

7. Obtain the probability of accomplishing the design project on the stated date using the formula:

$$w = \frac{T - T_L}{\left| \sum S_{cr}^2 \right|^{1/2}}$$

where

S_{cr}^2 = the variance of activities on the critical path

T_L = the last activity's earliest expected completion time, as calculated though the network

T = the design project due date, expressed in time units

Table 2.1 presents the probabilities for selective values of w.

TABLE 2.1 Probability Table

w	Probability
−3.0	0.0013
−2.5	0.006
−2.0	0.023
−1.5	0.067
−1.0	0.159
0.5	0.309
0.0	0.5
0.5	0.69
1.0	0.84
1.5	0.933
2.0	0.977
2.5	0.994
3.0	0.999

EXAMPLE 2.1[3]

A mechanical design project was broken down into a number of major jobs or
activities, as shown in Table 2.2. A CPM network was developed using the data
given in the table, and the critical path of the network was determined along with
the expected project duration time period.

Using the defined symbols and the data given in Table 2.2, the CPM network
is shown in Figure 2.3.

The paths originated at event 1 and terminated at 11 are as follows:

1. A-B-C-D-E-G-I-J time _20
2. A-B-C-D-F-G-I-J time _21
3. A-B-C-D-E-G-H-J time_18
4. A-B-C-D-F-G-H-J time _19.

These results indicate that the longest path is (A-B-C-D-F-G-I-J), and the predicted time for the design project is 21 weeks.

TABLE 2.2 Design Project

Activity Description	Activity Identification	Immediate Predecessor Activity or Activities	Activity Duration (Week or Weeks)
Needs, goals and market analysis	A	—	1
Function analysis	B	A	2
Specifications	C	B	1
Alternatives	D	C	4
Evaluations	E	D	3
Prototyping	F	D	4
Analysis	G	E,F	2
	H	G	2
Manufacturing	I	G	4
Marketing	J	H, I, 3	—

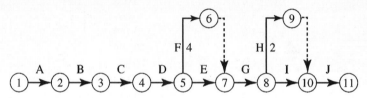

Figure 2.3 CPM chart.

[3]This example is based on an example from *Engineering Design—A Modern Approach*, by B.S.
Dhillon, Irwin Publishers.

LAB 5: Project Management (Microsoft Project)[4]

Purpose

In this lab we will introduce some main concepts of Project Management, and some techniques to control and monitor a project by using Microsoft Project 2007 software.

Management: The process of getting things done through the effort of other people by planning, organizing, and controlling these processes.

The Project: A complex effort usually less than three years in duration, made up of interrelated tasks performed by various organizations within well defined objectives, schedule, and budget.

The objective of the project management is to achieve proper control of the project to assure its completion on schedule and within budget, achieving the desired quality of the resulting product or services.

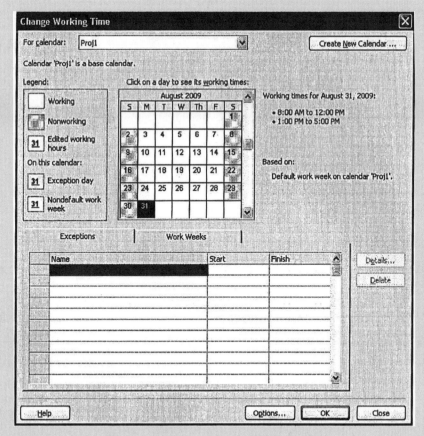

Proj1 calendar.

[4]© Dr. Adnan Bashir

Succeeding as a project manager requires that you complete your projects on time, finish within budget, and make sure your customers are happy with what you deliver. That sounds simple enough, but how many projects have you heard of (or worked on) that were completed late or cost too much or didn't meet the needs of their customers?

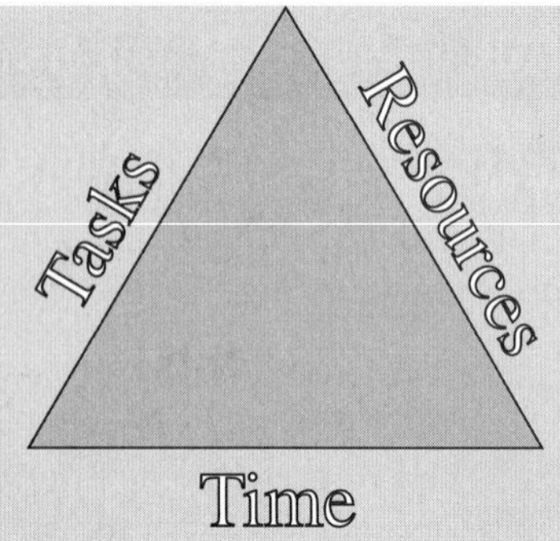

Figure L5.1   The project triangle

The Project Triangle: Seeing Projects in Terms of Time, Cost, and Scope

This theme has many variations, but the basic idea is that every project has some element of a time constraint, has some type of budget, and requires some amount of work to complete. (In other words, it has a defined scope.) The term *constraint* has a specific meaning in Microsoft Project, but here we're using the more general meaning of a limiting factor. Let's consider these constraints one at a time.

Time

Have you ever worked on a project that had a deadline? (Maybe we should ask, have you ever worked on a project that did not have a deadline?) Limited time is the one constraint of any project with which we are all probably most familiar. If you're working on a project right now, ask your team member what the project deadline is. They might not know the project budget or the scope of work in great detail, but chances are they all know the project deadline.

Cost

You might think of cost just as dollars, but project cost has a broader meaning: Costs are all the resources required to carry out the project. Cost includes the people and equipment who do the work, the materials they use, and all the other events and issues that require money or someone's attention in a project.

Scope

You should consider two aspects of scope: product scope and project scope. Every successful project produces a unique product: a tangible item or a service. You might develop some products for one customer you know by name. You might develop other products for millions of potential customers waiting to buy them (you hope). Customers usually have some expectations about the features and functions of products they consider purchasing. Product scope describes the intended quality, features, and functions of the product—often in minute detail. Documents that outline this information are sometimes called product specifications. A service or an event usually has some expected features as well. We all have expectations about what we'll do or see at a party, a concert, or the Olympic Games.

Project scope, on the other hand, describes the work required to deliver a product or a service with the intended product scope. Although product scope focuses on the customer or the user of the product, project scope is mainly the measure in *tasks* and *phases*.

A Typical Project Life Cycle

All projects can be described using a four-phase lifecycle, as shown in Figure L5.2.

In the first phase, a need is identified by the client, customer, or other stakeholder. This results in a process to describe and define the needs and requirements, sometimes soliciting information or proposals from vendors, contractors, or consultants. We can call this phase **initiation**.

The second phase is characterized by the development of **proposed solutions**. This can be by a structured bid from which requests are specific items of information related to project costs, staffing, timescales, description of the activities, compliance to technical standards, and key deliverables.

The third phase is when the project is actually executed covering detailed **planning** and **implementation**.

The final phase is terminating the project or **closure**. In some cases this is marked with formal acceptance by the customer or client with signed documentation.

Figure L5.3 illustrates how the concept of project phases is incorporated into a new product development methodology.

Reasons for Project Planning

- To eliminate or reduce the uncertainty.
- Improve efficiency of the operation.
- Obtain a better understanding of the project objectives.
- Provide a basis for monitoring and controlling work.
- Establish directions for project team.
- Support objectives of parent organization.
- Make allowance for risk.
- Put controls on the planned work.

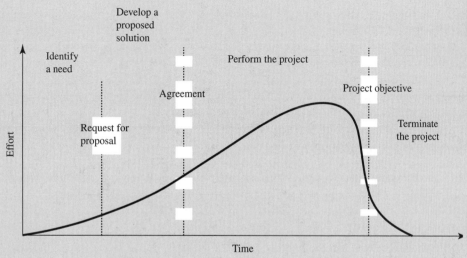

Figure L5.2 Four phase project cycle.

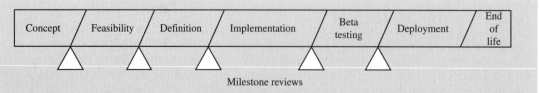

Milestone reviews

Figure L5.3

If the task is not well understood, then during the actual task execution more knowledge is learned, which leads to change in resource allocation schedules and properties. While if the task is well-understood prior to be performed, then work can be preplanned.

Project Scheduling

Project scheduling involves sequencing and allotting time to all project activities. Managers decide how long each activity will take and compute the resources (manpower, equipment, and materials) needed for each activity.

One popular project scheduling approach is the Gantt chart. **Gantt charts** are planning charts used to schedule resources and allocate time. Gantt charts are an example of a widely used, nonmathematical technique that is very popular with managers because it is so simple and visual. The major discrepancy with Gantt charts is the inability to show the interdependencies between events and activities.

The purposes of project scheduling are the following.

- Shows the relationship of each activity to others and to the whole project.
- Identifies the precedence relationships among activities.
- Encourages the setting of realistic time and cost estimates for each activity.
- Helps make better use of people, money, and material resources by identifying critical bottlenecks in the project.

Interdependencies are shown through the construction of networks. Network analysis can provide valuable information for planning, scheduling, and resource management. Program Evaluation and Review Technique (PERT) and Critical Path Method (CPM) are very popular and widely used network techniques for large and complex projects.

PERT: A technique to enable managers to schedule, monitor, and control large and complex projects by employing three time estimates for each activity.

CPM: A network technique using only one time factor per activity that enables managers to schedule, monitor, and control large and complex projects.

PERT and CPM were developed in 1958 and 1959, respectively, and spread rapidly throughout all industries. Mainly, CPM was concentrated in the construction and process industries.

Framework of PERT and CPM

- Define the project and prepare WBS.
- Develop the relationship between the activities.

- Draw the network connecting all the activities.
- Assign time/cost estimate to each activity.
- Compute the longest path through the network (**critical path**).
- Use the network to help plan, schedule, monitor, and control the project.

Example

Consider the following list of activities with its duration and predecessors. Find the critical path, and the draw the network, as in Figure L5.4.

Activity	Predecessor	Duration
a	—	5 days
b	—	4
c	a	3
d	a	4
e	a	6
f	b, c	4
g	d	5
h	d, e	6
i	f	6
j	g, h	4

Computerized PERT/CPM reports and charts are widely available today on personal computers. Some popular software are Primavera™ (by Primavera Systems, Inc.), MS Project™ (by Microsoft Corp.), MacProject™ (by Apple Computer Corp.), Pertmaster™ (by Westminster Software, Inc.), VisiSchedule™ (by Paladin Software Corp.), and Time Line™ (by Symantec Corp.). In this chapter we will use MS Project 2007™.

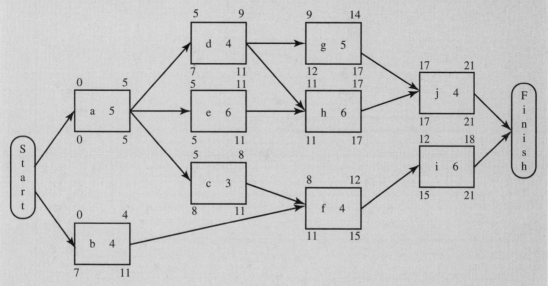

Figure L5.4 The critical path and time for the project is a-e-h-j with completion time of 21 days.

Using MS Project 2007

The first step to start the project is to build the project calendar that it will be used in the project. Figure L5.5 shows the standard calendar, which is the default in MS Project 2007.

We can create a new calendar to suit our project working days, vacations, number of shifts, etc. To create a new calendar, follow the following steps.

Go to tools
Go to change working time
Select Create New Calendar
Insert the new name of the calendar (proj1)
Choose copy from standard

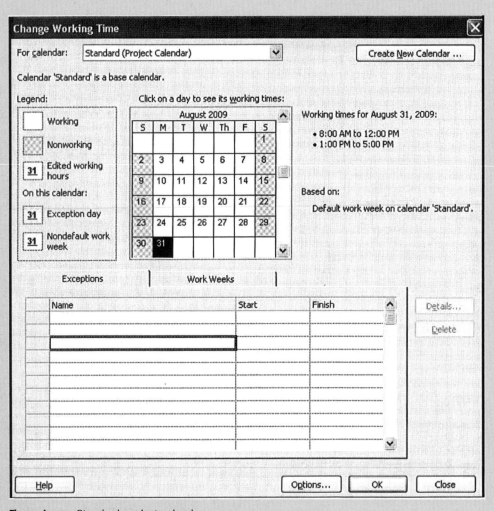

Figure L5.5   Standard project calendar.

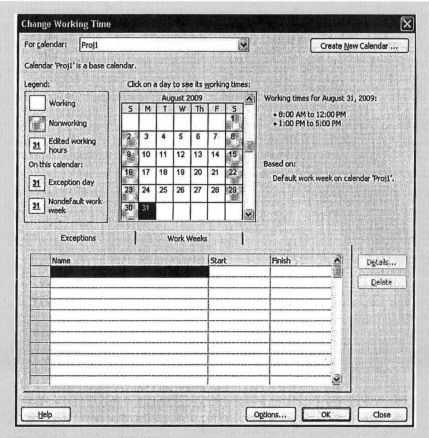

Figure L5.6 `Proj1` calendar.

You will get a new calendar called `proj1`, as shown in Figure L5.6. We can easily change the non-working days to Thursday and Friday by assigning them to nonworking time and assign the Saturday and Sunday to default (working days). Also, we can increase or decrease the number of working hours in a day by using non-default working time and fill in for propitiate working hours. Figure L5.7 shows the new calendar (`proj1`).

Entering Tasks and Subtasks

When you have identified the majority of the tasks in the project, preferably in a WBS, you should enter them into Microsoft Project. Do not worry about putting the tasks in the exact correct order in the beginning. It is easy to re-organize them later on.

You enter a task by one of the following ways.

1. Click the correct row in the Task Name column and enter the name of the Task.
2. Write the name of the Task in the Entry Bar. Click the correct row in the Task Name column and press the green check box next to the Entry Bar.

After you have entered the major tasks of the project it is time to add the details in terms of Subtasks. When adding subtasks the upper level task becomes a Summary Task. This is a good way

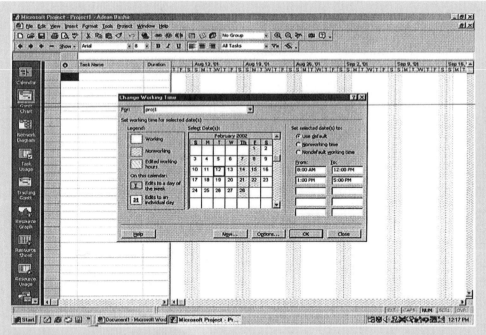

Figure L5.7   Creating a new calendar.

of structuring the project and displaying information easily. You are able to have up to nine levels of tasks.

You add a Subtask (Figure L5.8) using the following steps.

1. Click the row below the task that you would like to become the Summary Task in the Task Name column.
2. Write the name of the task and create a new task.
3. Click on the ID number of the row to mark the entire row.
4. Right-click on the row and choose Indent.

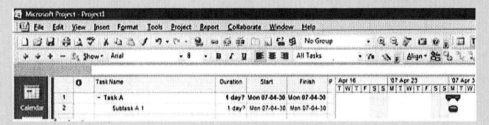

Figure L5.8   The summary task is in Bold and the Subtask is Indent. The summary task is displayed as a black bar on the Gantt chart.

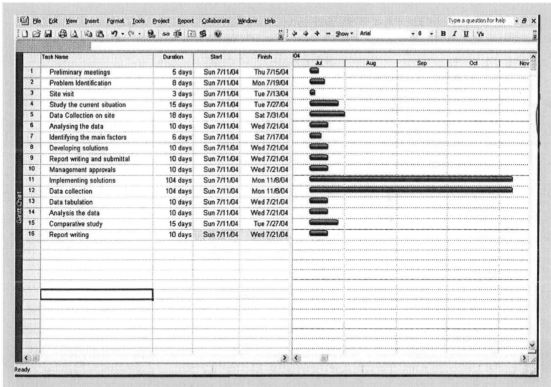

Figure L5.9   Task name and duration.

Once you setup the calendar to your project, you easily can now enter the tasks and the duration of your project, as shown in Figure L5.9

Once the task name and duration are entered, we can now insert the relation between the tasks by inserting a precedence column, and then we enter the task ID to identify the task. Figure L5.10 shows the task, duration, and the relationship between the tasks. The first activity shows a task summary with the duration covering the whole subtasks. The arrows between the bars represent the relation between the tasks—in this case the relationship is Finish to Start (FS); it means that the activity can not start until the preceding activity is completed.

Figure L5.10 shows the Gantt chart (scaled time bar chart for each task), task name, task duration, start, finish, and precedence. If we did not specify the start date of the project, MS Project 2007 will consider the current date as the project start date (as default). This Gantt chart is commonly used for most of the projects because it is simple, easy to understand, and easy to use to extract a summary information for the project. We can add any information to appear in the Gantt chart, such as Resources, Float (Slack), Percentage completion, Actual start, Actual finish, Late start, Late finish, Actual cost, Baseline cost, Milestones, Remaining duration, Remaining cost, etc.

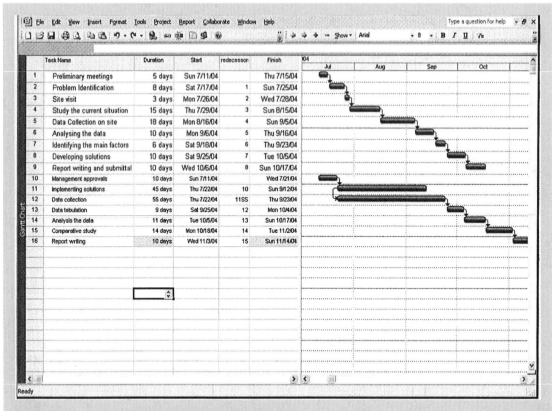

Figure L5.10   Tasks with their duration and relationships.

All of these can be added by

Go to insert
Select column
Pick the type of information you want to add
Press ok

If you double click on any task, a new window will appear. This window contains the following sub-windows:

General: Gives the name, duration, the percentage completion, Start and Finish dates of that task.

Predecessors: All the predecessors of this task will appear under this window showing its ID, Name, Type, and the Lag ID available.

Resources: You can assign the resources required to complete the task by entering the resource name and how many units needed.

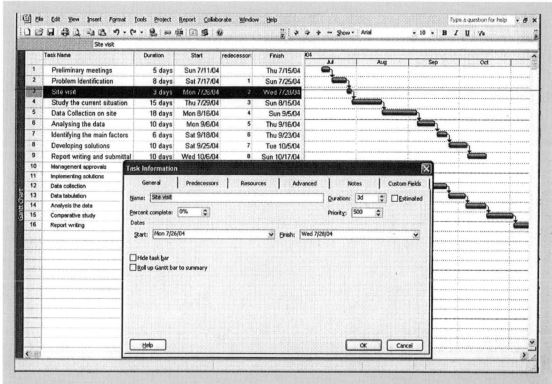

Figure L5.11   Task information window.

Advanced: This will give us the ability to insert time constraints, the task type, and the calendar. It shows also the Work Break Structure of the task (WBC).

Figure L5.11 shows the task information window for the third task. This task information window consists of the following buttons, each button contains full details of that button:

- General
- Predecessors
- Resources
- Advanced
- Notes
- Custom fields

To enter the resources of the project the following should be followed:

1. Go to resource sheet by clicking the view icon on the main menu and start to fill that sheet, as shown in Figure L5.12.
2. Assigning the resources to each task, as shown in Figure L5.13

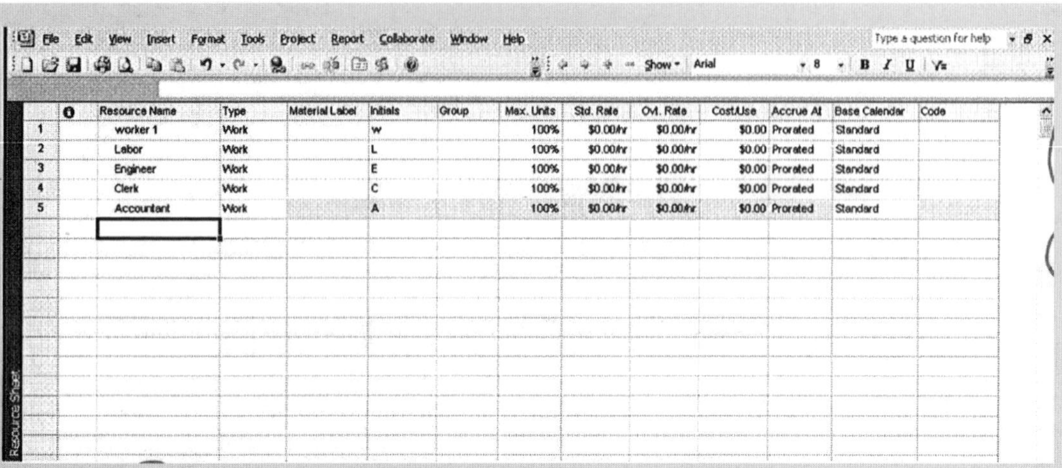

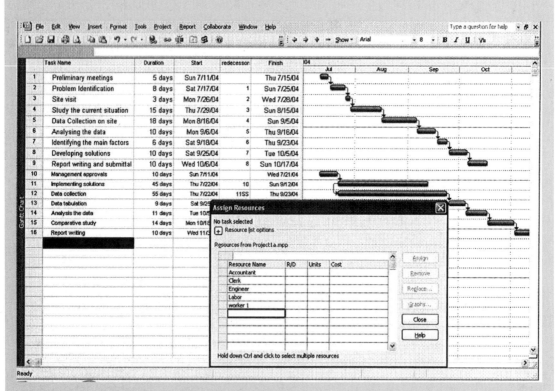

Figure L5.12   Resource sheet.

Figure L5.13   Assign task resources.

Go to tools
Resources, Assign resources
Pick the resource and units for that task

There is another way to assign the resources to tasks;

Once you define the resources
Go to Gantt chart
Insert new column "Resource"
For each single task enter the resources needed to complete that task

You are also able to see the planned workload for the different resources by clicking the Graphs button. You will then see the following the dialog box (Figure L5.14).

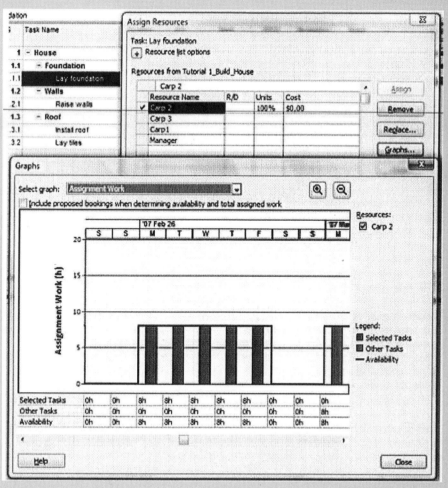

Figure L5.14   Graphs Dialog Box.

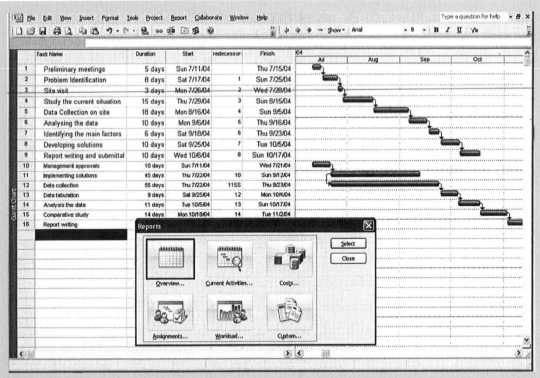

Figure L5.15   Types of reports in MS Project 2007.

The gray bars in the Figure L5.15 indicates the selected task and the blue bar to the far right shows another task. This graph is useful if you are uncertain if a specific resource is available for the task or not.

Reports in MS Project 2007

In order to get the output and the reports, MS Project 2007 provides several predefined reports, as shown in Figure L5.15.

Also, there is an option to customize any report you like, for example,

Go to tool bar, click Reports, and click Custom

Choose the type of report you need, then copy button

In the name box, type Custom Task Report

In the period box select Months from the drop-down list

In the table box select Summary from the drop-down list

Press ok (Figure L5.16 on the next page)

Click ok to close the Task Report dialog box

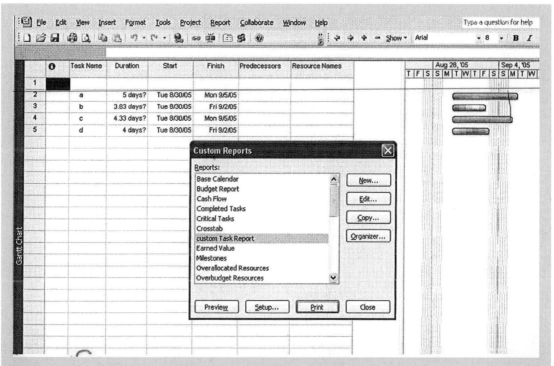

Figure L5.16   Customize a Task Report.

In the Custom Reports dialog box, click Preview
MS Project applies the custom report settings you choose, and the report appears
In the print preview window (Figure L5.17)

This custom report shows the fields displayed on the Summary Task table but dividing the tasks by month.

Updating Start and Finish Date

Remember that this information is essential for you as a project manager, and you should always be aware if you are on track or running late in the project. Not knowing the current status of the project will force you into making incorrect decisions and in the long run also hurt your project management career.

You update the actual progress (Figure L5.18) of a task by using the following steps.

1. Open the Gantt chart.
2. Choose Window->Split.
3. Click on the bottom pane.
4. Choose View->More Views . . .->Task Detail Form.

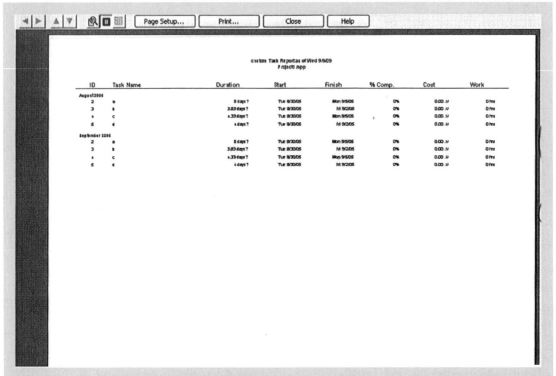

Figure L5.17   Customize a Summary Task Report.

5. Choose the task in the Gantt chart that you would like to update by clicking it.
6. Choose Actual in the Dates section.
7. Edit either Start or Finish date.
8. Confirm by clicking OK.

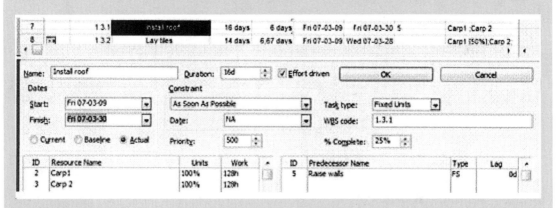

Figure L5.18   Update Actual Start or Finish date for the task Install Roof.

Example

Consider the following list of activities for building a small house project.

Activity	Description	Predecessor	Duration (days)
A	Clear site	—	1
B	Bring utilities to site	—	2
C	Excavate	A	1
D	Pour Foundation	C	2
E	Outside Plumbing	B,C	6
F	Frame House	D	10
G	Electric Wiring	F	3
H	Lay Floor	G	1
I	Lay Roof	F	1
J	Inside Plumbing	E,H	5
K	Shining	I	2
L	Outside sheathing insulation	F,J	1
M	Install windows / outside doors	F	2
N	Brick work	L, M	4
O	Install walls & ceiling	G,J	2
P	Cover walls & ceiling	O	2
Q	Install roof	I,P	1
R	Finish interior	P	7
S	Finish exterior	I,N	7
T	Landscape	S	3

- Determine the ES, LS, EF, and LF.
- Determine the Total Float.
- Determine the critical path and the total project duration.

Solution

- For this example, we will keep the default calendar (Standard).
- The project will start on 9/1/09. To enter this information:
    - *Go to project*
    - *Select project information, and put the start date of the project*
- Enter the task name, duration, and the predecessor for each task in MS Project software.
- Put the grid line for the major and minor columns:
    - *Click the right button of the mouse while the curser is in the bar chart area*
    - *Select the grid option*
    - *Select the major column and choose the suitable type of line, and its color*
    - *The same for the minor column*
- In order to get a full view of the Gantt chart:
    - *Go to view, and select zoom*
    - *Select the option "entire project"*

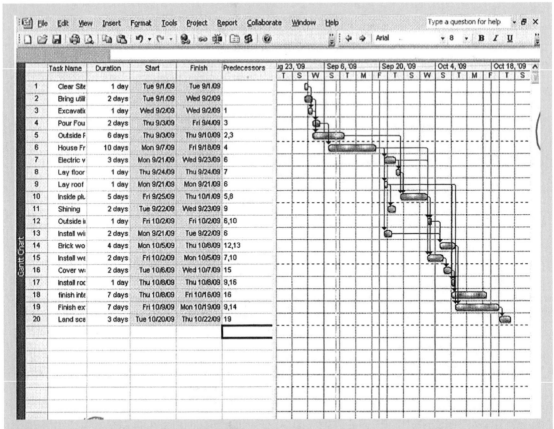

Figure L5.19   Gantt chart for Lab 5 example.

The outputs of these steps are shown in Figure L5.19.

- In order to find the ES, EF, LS, LF, and the Total Float:
    Go to insert
    Select column
    From the field name select Early Start, and press OK
    Repeat the same for EF, LS, LF, and Total Float
- To show the critical path and the critical activities:
    Go to format
    Select bar style
    Click insert row, and type Critical Activity
    Select the shape of the bar, and its color (red)
    In the field show for task, select critical, press OK
    Note that the critical tasks should have Total Slack = 0

The outputs are shown in Figure L5.20 on the next page.

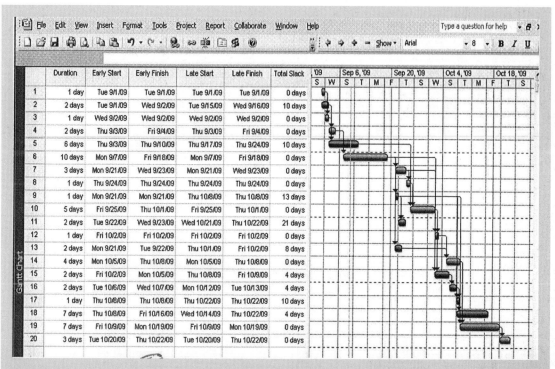

Figure L5.20 Gantt chart with ES, EF, LS, LF, and Total Slack.

- The total project duration can be obtained as follows:
  Go to Project
  Select project information
  Select statistics, and there will be much more information about the project as shown in Figure L5.21
- To show the network diagram:
  Go to view
  Select Network Diagram
  Use zoom in & zoom out to have the desired view

Figure L5.22 on page 81 shows the network diagram for the previous example, as well as the critical path on the network with the red color. Each node (box) contains: task name, duration, start date, finish date, and resources. Each node is connected to another node according to the relationship between the tasks.

Lab 5 Problems

1. What are the differences between Gantt, PERT and CPM?
2. What is the dummy activity, and why do we use it?

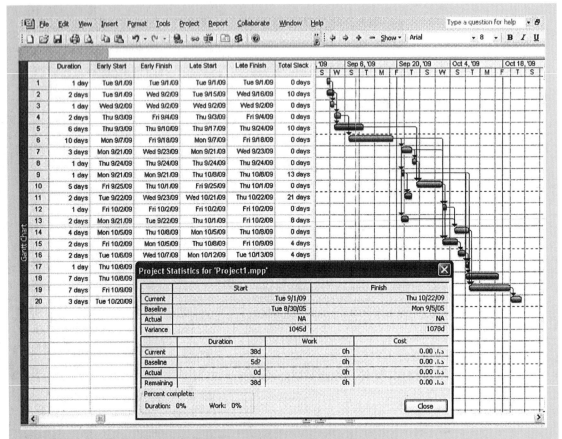

Figure L5.21   Project information statistics.

3. Why is planning important in projects?

4. What are the main reasons for project scheduling?

5. Construct a project network consisting of 12 activities (from 10 to 120) with the following relationships:
   a. 10, 20, and 30 are the first activities and are begun simultaneously.
   b. 10 and 20 must precede 40.
   c. 20 precedes 50 and 60.
   d. 30 and 60 must precede 70.
   e. 50 must precede 80 and 100.
   f. 40 and 80 must precede 90.
   g. 70 must precede 110 and 120.
   h. 90 must precede 120.
   i. 100, 110, and 120 are terminal activities.

6. Draw the CPM network associated with the following activities for a certain project. How long will it take to complete the project? What are the critical activities?

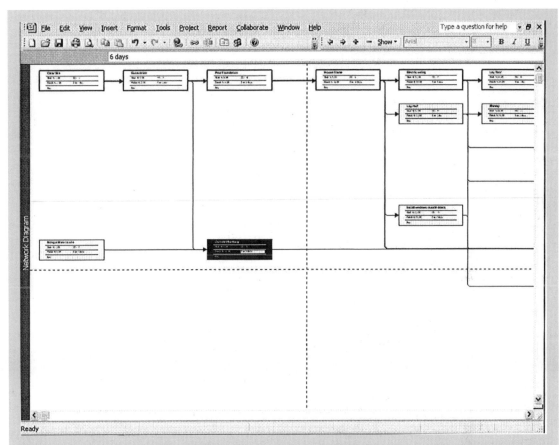

Figure L5.22 Part of the network diagram for the Lab 5 example.

Activity	Predecessor(s)	Time (min.)
A	—	2
B	A	4
C	A	6
D	B	6
E	B	4
F	C	4
G	D	6
H	E, F	4

7. For Problem 2, find the ES, EF, LS, LF, and Total Float for each activity.

8. ABC Company has a small project with the below listed activities. The manager of the project has been very concerned with the amount of time it takes to complete the project. Some of his workers are very unreliable. A list of activities, their optimistic completion time, the most likely completion time, and the pessimistic completion time (all in days) are given in the following table. Determine the expected completion time and the variance for each activity. Draw the PERT network diagram.

Activity	a	m	b	Predecessor(s)
A	3	6	8	—
B	2	4	4	—
C	1	2	3	—
D	6	7	8	C
E	2	4	6	B, D
F	6	10	14	A, E
G	1	2	4	A, E
H	3	6	9	F
I	10	11	12	G
J	14	16	20	C
K	2	8	10	H, I

9. In Problem 6, the project manager would like to determine the total project completion time and the critical path for the project. In addition, determine the ES, EF, LS, LF, and Total Slack for each activity.

10. In Problem 6, by using MS Project, produce a report showing the list of tasks that have an expected duration of more than five days.

11. Redo Problem 6 using the following calendar:
   a. Working hours per day is 10 hrs
   b. Working days per week is six days, (only Sunday is off)

References

HAROLD KERZNER, *Project Management: Systems Approach to Planning, Scheduling, and Controlling*, Ninth edition. John Wiley & Sons, Inc., NJ, 2006.

JAY HEIZER, BARRY RENDER, *Principles of Operations Management*, Fourth edition. Prentice Hall, NJ, 2001.

CARL CHATFIELD, TIMOTHY JOHNSON, *Step by Step Microsoft Office Project 2003*. Microsoft Press, WA, 2003.

MANTEL S. J., MEREDITH J. R., SHAFER S. M., and SUTTON M. M. *Project Management in Practice*, Third edition. John Wiley & Sons, Inc., NJ, 2008.

MEREDITH J. R. and MANTEL S. J., *Project Management—A Managerial Approach*, Sixth edition. John Wiley & Sons, Inc., NJ, 2006. ■

2.4   RESEARCH SKILLS

Research is the ability to collect data, gather the information, and interpret it as knowledge. This is a vital skill towards success and lifelong learning. Lifelong learning is the ability to keep developing and learning outside of the classroom. The ability to research allows a person to stay up-to-date with the latest in technology and the marketplace. Differentiating between data, information, and knowledge is a good start in appreciating what is required to attain and practice research skills. Data can be defined as raw, unprocessed material, which can be collected either through laboratory testing or many other sources. Once this data is processed into something meaningful, this becomes information. However, it only becomes knowledge when one is able to apply this information successfully.

Eisenburg and Berkowitz developed the Big6 and breaks down the research skills into the following six categories:

1. Task Definition
2. Information Seeking Strategies
3. Location and Access
4. Use of Information
5. Synthesis
6. Evaluation

The following is the interpretation of these points by the authors, however, for more detail, you may visit www.big6.com.

Step 1 defines the task and identifies what data and/or information needs to be collected or gathered. Once this is complete, the search for information can begin and step 2 provides a means to plan how you will go about seeking or collecting the information required. Step 3 identifies the sources (e.g. textbooks, internet, journals, magazines, etc.), and this is where you collect all the data and process it into information relevant to your task. Some material will have been processed already into information at the source, so all that is needed is the ability to be able to access and store it in an organized manner. Step 4 extracts the information that is needed, and Step 5 processes and presents this information into the final knowledge. Once knowledge has been attained, evaluation is key to ensuring that the knowledge and the process used to arrive at this knowledge is accurate and complete. In many cases, the knowledge may be accurate but incomplete and will open scope for further development.

Within the design process, research is utilized throughout all stages. However it is most evident immediately after identifying the customers' needs and goals. This is where the market analysis and information gathering stage for the project occurs. Chapter 4 discusses this in more detail.

2.5   TECHNICAL WRITING AND PRESENTATION

In the engineering and scientific professions, communication skills are as important as in other fields. Following the development of an innovative design and a cost estimate that predicts large profit, the designer must be able to communicate the findings to other people. An old adage reminds us that a tree falling in a forest doesn't make a sound unless there is someone to listen. Similarly, the best technical design in the world might never be implemented unless the designer can communicate the design to the proper people in the right way.

The quality of a report generally provides an image in the reader's mind that, in large measure, determines the reader's impression of the quality of work. Of course, an excellent job of report writing cannot disguise a sloppy investigation, but many excellent design studies have not received proper attention and credit because the work was reported in a careless manner. A formal technical report is usually written at the end of a project. Generally, it is a complete, stand-alone document written for people who have widely diverse backgrounds. Therefore, much more detail is required. The outline of a typical formal report is:

1. *Cover page:*
   a. Title of the report
   b. Name of author(s)
   c. Address
2. *Summary:* Summarizes the work and should also include a short conclusion. This may be the only section some people read (including the executives that may sponsor the project). Therefore, this section needs to stand alone. Writing a technical report is different to writing a novel, where 'giving away the ending' at the beginning of the book would not acceptable. In the case of a technical report, 'the ending' in the summary is not only acceptable, but it is actually expected.
3. *Table of contents.*
4. *List of figures and list of tables*, including the corresponding page number.
5. *Introduction:* Background of the work (market analysis) to acquaint the reader with the problems with and purposes of carrying on the work.
6. *Design process:* Details of the procedure followed in the design process.
7. *Discussion:* This section should contain the comprehensive explanation of the results. The discussion may be divided into several subsections, such as
   a. Technical analysis (force requirement, speed, etc.)
   b. Details of the build-up of the artifact
   c. Equipment used
   d. Assembly and setup procedures
   e. Experimental setups and results
   f. Details of the final results
8. *Conclusions:* This section states in as concise a form as possible the conclusions that can be drawn from the study.
9. *References:* This section lists all of the documents to which the writers referred. The information on each document must be complete and must follow the same format throughout.
10. *Appendices:* The appendices include material deemed beyond the scope of the main body of the report. The appendix section may be divided into as many subsections as necessary.

2.5.1 Steps in Writing a Report

The five operations involved in the writing of a high-quality report are best remembered with the acronym POWER:

1. **P**: Plan the writing
2. **O**: Outline the report
3. **W**: Write

4. **E**: Edit
5. **R**: Rewrite

Good writers usually produce good reports. Some of the attributes of good writing are:

1. Write as objectively as possible. Do not become emotionally involved or attached to a problem or a solution.
2. Be reasonably methodical.
3. Record whatever is learned, and keep in mind that whatever work is performed must eventually be documented.
4. Always strive for clarity in writing, and keep in mind that the written material should be simple and straightforward.
5. Deliver the written material on time.

There are also certain qualities associated with good reports, such as the following:

1. Delivered on the due date.
2. Effectively answer readers' questions as they arise.
3. Give a good first impression.
4. Read coherently.
5. Contain an effective summary and conclusion.
6. Are written clearly and concisely and avoid vague or superfluous phrases.
7. Provide pertinent information.

2.5.2 Illustration Guidelines

Visual elements (such as figures, charts, and graphs) in technical reports have specific purposes and convey specific information. Visuals are used to explain, illustrate, demonstrate, verify, or support written material. Visuals are only valuable if their presentation is effective. The general guidelines for preparing effective visuals (illustrations) are

1. Reference all illustrations in the text.
2. Reference the data source.
3. Carefully plan the placement of illustrations.
4. Specify all units of measure and the scale used in the drawing.
5. Label each illustration with an identifying caption and title. Include the figure source if the figure was obtained from another document.
6. Spell words out rather than use abbreviations.
7. When a document has five or more figures, include a list of figures at the beginning of the report.
8. Avoid putting too much data in an illustration.
   Figures 2.4 and 2.5 demonstrate the same illustration. However. Figure 2.4 is how not to do it. In this figure, the labeling is missing. The scale does not make efficient use of displaying the curve. The thickness of the line and plot symbols are too big to

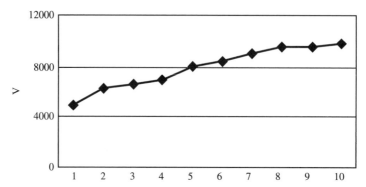

Figure 2.4   Example of a bad illustration.

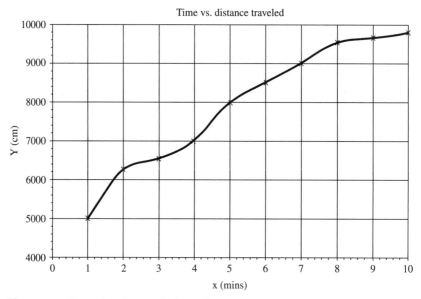

Figure 2.5   Example of a good illustration.

determine any form of accurate reading from the graph. The units and description of what the graph is measuring are also missing. The graph also seems to have been poorly scanned and has been displayed at a slight angle. Have a look at Figure 2.5 for a better example of the same illustration. Try to see what else has been improved.

2.5.3 Mechanics of Writing

The following suggestions will help you avoid some of the most common mistakes in your writing:

• *Paragraph structure:* Each paragraph should begin with a topic sentence that provides an overall understanding of the paragraph. Each paragraph should have a single theme or conclusion, and the topic sentence states that theme or conclusion.

- *Sentence length:* Sentences should be kept as short as possible so that their structure is simple and readable. Long sentences require complex construction, provide an abundance of opportunity for grammatical errors, take considerable writing time, and slow the reader. Long sentences are often the result of putting together two independent thoughts that could be stated better in separate sentences.
- *Pronouns:* There is no room for ambiguity between a pronoun and the noun for which it is used. Novices commonly use *it, this, that,* and so on where it would be better to use one of several nouns. It may be clear to the writer but is often ambiguous to the reader. In general, personal pronouns (*I, you, he, she, my, mine, our, us*) are not used in reports.
- *Spelling and punctuation:* Errors in these basic elements of writing are inexcusable in the final draft of the report.
- *Tense:* Use the following rules when choosing the verb tense:
  a. *Past tense:* Use to describe work done when you were building or designing or in general when referring to past events.
  b. *Present tense:* Use in reference to items and ideas in the report itself.
  c. *Future tense:* Use in making predictions from the data or results that will be applicable in the future.

2.6 PRESENTATION STYLE

Unlike advertising executives, engineers are ill equipped to sell their ideas. Secondhand information represented by company officials may not answer all of the client's questions. This section discusses the specifics and techniques of oral presentations.

2.6.1 Objective

Every presentation should have an objective. The speaker's main objective is to deliver the message (objective) to the audience. The objectives may vary from one presentation to another. To identify the real objective, ask the following question:

"If everything goes perfectly, what do I intend to achieve?"

Realize who your audience is and what their educational level may be.

In most cases, presentation time is limited. It is of utmost importance to keep within the scheduled time during question/answer sessions. To stay within time restraints, detailed planning is required. Different tools can be used in the presentation, such as slides, models, transparencies, audiovisual, and the Web. Make sure to account for the time needed to shift from one medium to another.

An easy way to evaluate the effectiveness of a presentation while on a team is to practice through role playing. One person can play the role of the speaker, and the rest of the team can act as the audience and possibly even play the devil's advocate. In this way, the team gains valuable experience before attempting the actual presentation.

2.6.2 Oral Presentation Obstacles

To sell your ideas to others, you should first be convinced that your ideas will accomplish the task at hand. Oral presentation requires a high degree of creativity.

People resist change, although they may announce that they embrace it. Humans like familiar methods. Change requires additional efforts, which humans, in general, resist. Typical reactions to change include

- We tried before.
- It is a too radical change.
- We have never done that before.
- Get back to reality.
- We have always done it this way.
- I don't like the idea.

Keep in mind the following:

- Only 70% of the spoken word is actually received and understood. Complete understanding can come through repetition and redundancy in speech.
- People mostly understand three-dimensional objects. Two-dimensional projections need to be transmitted with added details.
- People usually perceive problems from their own perspectives.
- Convey ideas so that they may be interpreted with the least expenditure of energy.

2.6.3 Oral Presentation Dos and Don'ts

Remember the following in making oral presentations:

1. Know your audience thoroughly.
2. Never read solely from notes, a sheet, or directly from an overhead projector. You can use your notes for reference, but remember to make eye contact with your audience from time to time.
3. Bring the audience up to the speed in the first few moments.
4. Stay within the time allotted.
5. Include relevant humorous stories, anecdotes, or jokes (only if you are good at it).
6. Avoid using specialized technical jargon. Explain the terms you feel the audience may not know.
7. Understand your message clearly. The entire goal is to communicate the message clearly.
8. Practice, practice, practice! You may like to memorize the introduction and concluding remarks.
9. The dry run is a dress rehearsal. Use it to iron out problems in delivery, organization, and timing.
10. Avoid mannerisms: Speak confidently but not aggressively.

11. Maintain eye contact with audience members, and keep shifting that contact through the talk.

12. Never talk to the board or to empty space.

13. Present the material in a clever fashion, but not in a cheap and sensational fashion. Be genuinely sincere and professional.

14. A logical presentation is much more critical in an oral one than it is in a written one.

2.6.4 Oral Presentation Techniques

The following will help you make your oral presentation as effective as possible:

1. Visual aids (sketches, graphs, drawings, photos, models, slides, transparencies, and the Web) often convey information more efficiently and effectively. Visual aids permit the use of both the hearing and seeing senses, and they help the speaker.

2. Limit slides to not more than one per minute.

3. Each slide should contain one idea.

4. The first slide should show the title of your presentation and the names of the collaborators.

5. The second slide should give a brief outline of the presentation.

6. The last slide should summarize the message you just delivered.

7. If you need to show a slide more than once, use a second copy.

8. Avoid leaving a slide on the screen if you have finished discussion on the topic.

9. Never read directly from the slide. Spoken words should complement the slides. Prepare notes for each slide and use them during practice.

10. Use graphs to explain variations. Clearly label the axis, data, and title. Acknowledge the source.

11. Every graph should have a message (idea). Color should enhance the communication, not distract from it.

12. Audiences respond to well-organized information. That means
    a. Efficient presentation
    b. All assumptions clearly stated and justified
    c. Sources of information and facts clearly outlined

13. Begin with the presentation of the problem and conclusion/recommendation (primary goal).

14. Finish ahead of schedule and be prepared for the question/answer session.

2.6.5 Question/Answer Session

The question/answer session is very important. It shows the enthusiasm of the audience and usually reveals interest and attention. In the Q/A session you should

1. Allow the questioner to complete the question before answering.

2. Avoid being argumentative.

3. Do not let the questioner feel that the question is stupid.
4. Adjourn the meeting if the questions slack off.
5. Thank the audience one final time after the Q/A session.

LAB 6: Presentation Style

Purpose

Unlike the advertising executive, the engineer is ill equipped to sell his or her ideas. Second-hand information represented by company officials may not answer all the client's questions. This lab provides startup training for your presentation style.

Establish Objectives

Every presentation should have an objective. The speaker's main objective is to deliver the message (objective) to the audience. The objectives may vary from one speech to another. To identify the real objective, ask the following question: "If everything goes perfectly, what do I intend to achieve?" Also,

Even the best idea or design will not be effective if you cannot properly communicate your ideas to others. (Igor Karon/Shutterstock)

know your audience and their education level. In most cases, time is limited. It is of utmost importance to keep within the scheduled time for the question/answer session. Doing so requires detailed planning. Different audiovisual tools can be used in the presentation, such as slides, models, transparencies, and the Web. If more than one medium is used, allow for time to shift from one medium to another. One easy way to gauge the effectiveness of a presentation is for one team member to play the role of the speaker and the rest of the team to pose as the audience (a team member may also play the devil's advocate).

Oral Presentation Obstacles

To sell your ideas to others, you should first be convinced that your ideas will accomplish the task at hand. Oral presentation requires a high degree of creative ability. Humans resist change; they like familiar methods. Changes require additional effort, which humans in general resist. The following are typical reactions to proposed changes:

- We tried before.
- It is too radical a change.
- We have never done that before.
- Get back to reality.
- We have always done it this way.
- I don't like the idea.

Keep in mind that

- Only 70% of spoken words are actually received and understood. Complete understanding can come through repetition and redundancy in speech.
- People best understand three-dimensional objects. Two-dimensional projections need to be transmitted with added details.
- When considering a proposed change, people usually perceive potential problems.
- Convey ideas so that they are interpreted with the least expenditure of energy.

Oral Presentation Dos and Don'ts

1. Know your audience thoroughly.
2. Never read directly from notes, a sheet, or directly from an overhead projector.
3. Bring your audience up to speed in the first few moments.
4. Stay within the time allotted.
5. Include relevant humorous stories, anecdotes, or jokes (only if you are good at it).
6. Avoid using specialized technical jargon. Explain the terms if you feel the audience may not know them.
7. State your message clearly.
8. Practice, practice, practice. You may want to memorize the introduction and concluding remarks.

9. The dry run is a dress rehearsal. Use it to iron out problems in delivery, organization, and timing.
10. Avoid mannerisms: Speak confidently but not aggressively.
11. Maintain eye contact with some audience members and keep shifting that contact throughout the presentation.
12. Never talk to the board or to empty space.
13. Present the material in a clever fashion, but not in a cheap and sensational fashion. Be genuinely sincere and professional.
14. Logical presentation is much more critical in oral than in written presentation.

Oral Presentation Techniques

1. Visual aids (sketches, graphs, drawings, photos, models, slides, transparencies, the Web) often convey information efficiently and effectively. Visual aids permit a dual sense of hearing and seeing, and they help the speaker.
2. Limit slides to not more than one per minute.
3. Each slide should contain one idea and text that is clearly readable by the audience sitting furthest away from you.
4. The first slide should show the title of your presentation and names of collaborators.
5. The second slide should give a brief outline of the presentation.
6. The last slide should summarize the message you just delivered.
7. If you need to show a slide more than once, use a second copy.
8. Avoid leaving a slide on the screen if you have finished discussion on that topic.
9. Never read directly from the slide. Spoken words should complement the slides. Prepare notes for each slide and use them during practice.
10. Use graphs to explain variations. Clearly label the axis, data, and title. Acknowledge the source.
11. Every graph should have a message (idea). Color should enhance the communication, not distract.
12. Audiences respond to well-organized information. That includes
    a. Efficient presentation
    b. All assumptions clearly stated and justified
    c. Sources of information and facts clearly outlined
13. Begin with the presentation of the problem and conclusion/recommendation (primary goal).
14. Finish ahead of time and be prepared for the question/answer session.

Question/Answer Session

The question/answer session is very important. It shows the audience's enthusiasm, interest, and attention. In the Q/A session you should

1. Allow the questioner to complete the question before answering.
2. Avoid being argumentative.
3. Do not let the questioner feel that the question is stupid.
4. Adjourn the meeting if the questions slack off.
5. Thank the audience one final time after the Q/A session.

Lab 6 Problems

Choose one of the following topics and prepare a 5-minute presentation for your classmate or teammates. Adhere to the instructions provided in this lab. Your classmates or teammates will provide anonymous evaluation scores through the instructor. The evaluation scores consists of the following elements:

1. Level of preparation (out of 10 points). What level of preparation was expressed in the slides' quality and content?
2. Level of audience engagement (out of 5 points). Did the speaker maintain eye contact with the audience?
3. Level of professionalism (out of 5 points). Was the speaker dressed properly, and did he or she act professionally during the presentation?
4. Communication skills (out of 5 points). Was the presentation clear? Was the speaker able to convey the message?
5. Time management (out of 5 points). Did the speaker leave time for questions? How much time is really needed to convey the message?
6. Overall performance (out of 10 points).

The instructor will collect the evaluations and pass them on to the presenter. The purpose of this exercise is to provide you with an evaluation of your presentation skills. You need to enhance the strong aspects and work on improving the weak aspects of presentation technique.

The following is a suggested list of topics. Students may choose a different topic with instructor consent.

1. How can a person maintain reading time in a busy schedule?
2. How can a school library raise funds for more resources?
3. How can engineers keep up with the latest technology?
4. How can you maintain a steady study schedule when you have tons of homework?
5. What are the pros and cons of teamwork?
6. How can teamwork become a nightmare?
7. How can teamwork become fruitful?
8. Why is design a social activity?
9. Compare the old versus the new paradigms of design. ■

2.7 PROBLEMS

2.7.1   Team Activities

Scheduling

1. Painting a two-story house was broken down into a number of major jobs or activities, as shown in Table 2.3.
   (a) Develop a CPM network.
   (b) Determine the critical path of the network.
   (c) Determine the expected project duration time period.
2. Assume that the optimistic, most likely, and pessimistic activity times are as given in Table 2.4.
   (a) Complete the table.
   (b) Find the probability of finishing the job in 32 weeks.
3. You should work on this activity during lab hours.
   (a) Develop a CPM network and determine the critical path for the events defined in the Gantt chart in Figure 2.1.
   (b) Develop a Gantt chart for the events defined in the Table 2.2.
   (c) For the events defined in Table 2.5,
       i. Complete the table.
       ii. Draw the PERT network.
       iii. Find the probability of finishing the task on time if the design due date is after 95 days.

2.7.2 Individual Activities

4. Define the following terms:
   (a) CPM
   (b) Gantt chart
   (c) PERT

TABLE 2.3 Team Problem 1

Activity	Identification	Predecessor	Duration
Contract signed	A	—	2
Purchase of material	B	A	2
Ladder & staging in site	C	A	2
Preparation of surface	D	C,B	5
Base coat complete	E	D	6
Base coat inspected	F	E	2
Trim coat complete	G	E	5
Trim coat inspected	H	G,F	2
Final inspection	I	H	2
Removal of staging	J	H	2
Final cleanup	K	I,J	2

TABLE 2.4 Team Problem 2

Pessimistic	Most Likely	Optimistic	Expected Time	Variance
3	2	1		
3	2	1		
3	2	1		
7	5	3		
9	6	4		
3	2	1		
7	5	4		
3	2	1		
3	2	1		
3	2	1		
3	2	1		

TABLE 2.5 Team Problem 3

Predecessor	Optimistic	Most Likely	Pessimistic	Expected	Variance
1	3	5	8		
1	4	6	9		
3	3	4	5		
2	2	3	4		
5	3	4	5		
1	8	12	14		
4	14	18	21		
7	5	10	14		
7	5	10	14		
7	5	10	14		
7	5	10	14		
7	5	10	14		
12	4	6	10		
11	4	6	10		
10	4	6	10		
9	4	6	10		
8	4	6	10		
13	10	12	18		
6	16	18	24		
14	7	10	15		
14	10	15	22		
17	5	9	9		
18	4	6	8		
16	6	8	12		
15	3	8	12		
19	3	4	5		

TABLE 2.6 Term Project

Activity Description	Activity Identification	Immediate Predecessor Activity	Activity Duration (Days)
Literature collection	A	—	7
Literature review	B	A	4
Outline preparation	C	B	1
Analysis	D	B	10
Report writing	E	C	5
Typing	F	E	3
Revision	G	E	4
Final draft	H	G	2

5. Assume that an engineering course term project is broken down into a number of major jobs or activities, as shown in Table 2.6.
   (a) Draw a Gantt chart.
   (b) Develop a CPM network.
   (c) Determine the critical path of the network.
   (d) Determine the project duration time period.
6. Assume that, for example, the optimistic, most likely, and pessimistic activity times are as in Table 2.7. Calculate each activity's expected time and variance, and the probability of accomplishing the design project in 32.5 weeks. In addition, calculate each event's earliest and latest event times.

TABLE 2.7 Activity Time Estimates (Weeks)

Pessimistic	Most Likely	Optimistic	Expected Time	Variance
4	2	1	2.16	0.25
3	2	1	2	0.11
4	2	1	2.16	0.25
8	5	3	5.16	0.7
5	4	2	3.83	0.25
7	4	3	4.33	0.44
5	3	2	3.17	0.25
5	3	2	3.17	0.25
9	6	3	6	1
7	4	2	4.17	0.69

2.8   Selected Bibliography

BEARD, P.D., and TALBOT, T. F. "What Determines If a Design Is Safe?." *Proceedings of ASME Winter Annual Meeting*, pp. 90–WA/DE-20, New York, 1990.

BLAKE, A. *Practical Stress Analysis in Engineering Design*. New York: Marcel Dekker, 1982.

BURGESS, J.H. *Designing for Humans: The Human Factor in Engineering*. Princeton, NJ: Petrocelli Books, 1986.

BURR, A. C. *Mechanical Analysis and Design*. New York: Elsevier Science, 1962.

COLLINS, J.A. *Failure of Materials in Mechanical Design*. New York: Wiley, 1981.

COOK, N. H. *Mechanics and Materials for Design*. New York: McGraw-Hill, 1984.

CULLUM, R.D. *Handbook of Engineering Design*. London: Butterworth, 1988.

DEUTSCHMAN, A.D. *Machine Design: Theory and Practice*. New York: Macmillan, 1975.

DHILLON, B. S. *Engineering Design: A Modern Approach*. Toronto: Irwin, 1996.

DIESCH, K. H. *Analytical Methods in Project Management*. Ames, Iowa: Iowa State University, 1987.

DIETER, G. E. *Engineering Design: A Material and Processing Approach*. New York: McGraw-Hill, 1983.

DOYLE, L. E. *Manufacturing Processes and Material for Engineers*. Englewood Cliffs, NJ: Prentice Hall, 1985.

DREYFUSS, H. *The Measure of Man: Human Factors in Design*. New York: Whitney Library of Design, 1967.

EISENBERG, M. B. and Berkowitz, R. E. Big6™. http://www.big6.com.

FREDERICK, S.W. "Human Energy in Manual Lifting." *Modern Materials Handling*, Vol. 14, pp. 74–76, 1959.

FURMAN, T.T. *Approximate Methods in Engineering Design*. New York: Academic Press, 1980.

GLEGG, G. L. *The Development of Design*. Cambridge, UK: Cambridge University Press, 1981.

GRANDJEAN, E. *Fitting the Task of the Man: An Ergonomic Approach*. London: Taylor and Francis, 1980.

GREENWOOD, D. C. *Engineering Data for Product Design*. New York: McGraw-Hill, 1961.

HAJEK, V. *Management of Engineering Projects*. New York: McGraw-Hill, 1984.

HILTON, J. R. *Design Engineering Project Management: A Reference*. Lancaster, PA: Technomic Publishing Co., 1985.

HINDHEDE, A. *Machine Design Fundamentals*. New York: Wiley, 1983.

HUBKA, V., and EDER, W. E. *Principles of Engineering Design*. London: Butterworth Scientific, 1982.

LUPTON, T. *Human Factors: Man, Machine and New Technology*. New York: Springer- Verlag, 1986.

McCORMICK, E. J. and SANDERS, M. S. *Human Factors in Engineering and Design*. New York: McGraw-Hill, 1982.

MEREDITH, D.D. *Design and Planning of Engineering Systems*. Englewood Cliffs, NJ: Prentice Hall, 1985.

MOHRMAN S. A. and MOHRMAN A. M. *Designing and Leading Team-Based Organization: A Workbook for Organizational Self Design*. San Francisco: Jossey-Bass 1977.

OBORNE, D. J. *Ergonomics at Work*. New York: Wiley, 1982.

PAPANEK, V. *Design for Human Scale*. New York: Van Nostrand Reinhold, 1983.

PHEASANT, S.T. *Bodyspace: Anthropometry, Ergonomics and Design*. London: Taylor and Francis, 1986.

SCHMIDTKE, H. *Ergonomic Data for Equipment Design*. New York: Plenum Press, 1985.

SHIGLEY, J. E., AND MISCHKE, C. R. *Mechanical Engineering Design*. New York: McGraw-Hill, 1983.

WOODSON, W. E. *Human Factors Design Handbook*. New York: McGraw-Hill, 1981.

CHAPTER •3

Identifying Needs and Gathering Information (Market Analysis)

A market analysis is required at the start of the design process. Designers will use this market analysis to identify their market and fine tune their designs. (Wrangler/Shutterstock)

3.1 OBJECTIVES

By the end of this chapter, you should be able to

1. Identify and abstract the statement of need.
2. Identify data sources for your search.
3. Further clarify the statement of need.
4. Identify Standard Industrial Classification (SIC) codes for the industry.
5. Find what is available in the market that may provide you with leads to produce a better product.
6. Conduct market surveys and market analysis for your product development.

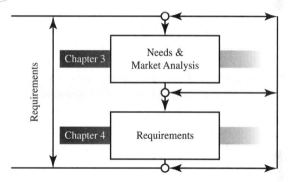

In general, need statements are not well defined. Even if they are, you will always need to make sure that what you are developing is what the customer needs. The mechanism of clarifying the need statement is discussed in this chapter and the following three chapters. This chapter discusses identifying and clarifying the customers' needs by searching what is already on the market and by knowing who you are competing with. Chapter 4 discusses organizing and prioritizing customer needs to match your understanding. Chapter 5 provides the tools to help you shape your ideas to solve the problem. Chapter 6 discusses revising objectives from qualitative definitions into quantitative definitions. In Chapter 7, through this systematic process, you begin to develop definitive solutions.

3.2 PROBLEM DEFINITION: NEED STATEMENT

During the course of human development, different kinds of needs existed. For instance, there has always been and always will be a need for improving and making new designs. Lincoln Steffens wrote. "The world is yours, nothing is done, nothing is known. The greatest poem isn't written, the best railroad isn't built yet, the perfect state hasn't been thought of. Everything remains to be done right, everything." The engineer is a person who applies scientific knowledge to satisfy humankind's needs. It should be emphasized that the ability to design is a characteristic of an engineer.

One serious difficulty that engineers must overcome deals with the form in which problems are often presented to them. Even if some goals are given to the engineer, they often are not specifically stated. Problems may be presented vaguely: "The shaft is breaking." "The controls aren't producing the desired effect." "It costs too much to operate this engine." Thus, the first task of the engineer involves determining the real problems. Then, the engineer must determine the extent and confines of the goals.

It is necessary to formulate a clear, exact statement of the problem in engineering words and symbols. It is also necessary to isolate the problem form the general situation and to delineate its form. This definition should clearly identify every aspect of the problem on which attention should be concentrated. The nonessential should be stripped away, and the individual characteristics of the problem should be differentiated. It should be determined whether or not the immediate problem is part of the larger problem. If it is, its relationship to the total part should be determined. Consider the following examples.

1. A designer is presented with a situation involving the waste of irrigation water in public parks. The park keepers forget to turn off the water. A general formulation of the problem would be "What can we do to minimize the possibility of workers forgetting to turn off the water before the end of their shift?" An engineer could ask the following questions. "Why do workers continue to forget to turn off the water?" "What is the sequence of events that workers use during their daily activities?" "What will happen if a keeper does not show up for his/her shift?" "Do we need to manually turn on and off the water?"

   A more precise form of the problem statement would be "How do we prevent irrigation water waste in public parks?"

2. A company has proposed to use the density gradient to isolate red blood cells from whole blood and thus to treat white blood cells with a light-activated drug. The designer should ask questions such as the following. "Is it necessary to use the density gradient if other methods of separation would be capable of isolating the red blood cells from the whole blood?" "If the white cells are being treated, why don't we isolate the white cells from the whole blood rather than isolate red blood cells?" "Why don't we impede the light into the blood and reduce the need for separation?"

   Designers need to abstract the need statement from its current state to a statement that they can base their design on. Vague statements from the customer usually result in a bad design.

3. An engineer is presented with a problem caused by the formation of ice on roofs. The ice forms during certain types of weather, falls away from the roofs, and causes damage to vehicles and people below. A general formulation of this problem might be "How do we prevent ice from forming on roofs?" However, further questions may be asked. "What would happen if ice did form?" "What will cause the ice to fall?" "What harm would such formation do?" These questions determine that the first definition was much too narrow. A much broader definition was "How do we prevent ice that forms on roofs from doing harm or damage to people and equipment below?"

   Before an engineer can define the problem properly, he or she must recognize all of the problems that exist. Most of the failures in machines do not occur because we make mistakes in analyzing the problem, but because we fail to recognize that there is a problem. For example, the first-year maintenance cost for a paint-drying oven is $2000 (an oven whose original cost is $1700). The high cost is due to the frequent dismantling of oil in bronze sleeve bearings mounted within the baking chamber. Once the problem is recognized, the solution is relatively simple (mount the bearings outside the oven).

4. A motor is used to drive a magnetically driven pump. Powerful magnets are attached at the shaft of the motor. The motor continues to fail because the magnets continue to interact with the stator magnets in the motor. Once the problem is identified, the solution is easy (put the magnets at a further location from the motor; you can use a belt system to drive the magnets).

So, it is evident that the needs should be identified clearly, otherwise a vague statement of need will lead to a vague understanding of the product to be designed. A vague understanding cannot give a solution that addresses the specific problem. Asking the right question requires engineering knowledge, practice, and common sense. There are also techniques which enable you to gather as much information as possible about the problem and needs of customers. These will be discussed throughout the remainder of this chapter.

3.3 GATHERING INFORMATION: CLARIFYING THE NEED

Identifying customer needs can be done in several ways. Either the designer is approached directly by a customer with a specific problem, or the designer finds an opportunity in the market by identifying a need for a new or improved product. In the design process, it is important to determine if such a business opportunity exists for a product before investing time and money to evolve the product. Most likely, products will fail when the market analysis only skims the surface. The analysis requires a thorough study of the total market. This includes its trend, competition, volume, profit, opportunities, consumer needs, and some indication of customer feeling for the product. Market analysis must be conducted at the start of the design process. It serves as a mechanism to define further the need statement and provides an opportunity for designers to review other attempts at solving the problem at hand, if they exist. Remember, the majority of designs are development designs. Thus, knowing what others have provided as a solution is very important before you attempt to offer your own solution.

It is also important at this stage not to provide solutions; you are gathering information with which to provide a better solution. This phase of the design process will help you appreciate your development and allow you to review what others have done in solving the same problem. It will also help you identify the size of the need and whom you are competing with. Furthermore, it will show what is available on the market.

King defines a market as a group of potential customers who have something in common. Two different techniques are employed in market analysis.

1. *Direct search:* This involves obtaining information directly from the consumer, manufacturers, salespeople, and so forth. The information is collected by interviews and surveys.

2. *Indirect search:* Information is collected from public sources, such as patents, journal reports, government analysis, and newspapers.

The market search should be done in a systematic and objective manner and consider all information that is relevant to the product. It is important to remember that an unbiased outlook is necessary when analyzing the data to formulate a market analysis report. Any biases will be reflected in the report's view of the market, and the results could be disastrous.

3.4   HOW TO CONDUCT A MARKET ANALYSIS[1]

Thousands of resources, it seems, all present some variation of the same information. However, it is never presented in the form you need. Before beginning any research project, it is important to plan the process. Information collection is usually one of the first steps, and traditionally, it was often anticipated with the same enthusiasm as a root canal. However, this is changing, as modern engineers have realized that it is an essential element of the design process if they wish to remain competitive in a global market. For the engineering student, product/market research often requires the use of unfamiliar information sources and often produces huge quantities of information. It is important that engineering skills be applied when planning and dealing with this deluge. These skills include

- Critical thinking
- Strategy
- Analysis
- Time management (it always takes longer than you think)

In the product development process, market research is conducted initially to assess market potential, market segments, and product opportunities and to provide production cost estimates and information on product cost, sales potential, industry trends, and customer needs and expectations. The search has the following steps:

1. Define the Problem
2. Develop a Strategy
3. Organize and check the information gathered

3.4.1   Define the Problem

Knowing what to look for is very important before you start gathering information. This can be accomplished by answering questions similar to the following.

a. Are you developing a new product or solving a problem in an existing product? *Remember, you are not providing a solution, you are redefining a problem.*

b. Who are your customers, and why would they want/need to buy the product (e.g., time saver, utility, unique value)?

c. What are the main needs of these customers?

---

[1]Information provided in this section is based on Suzanne Weiner's work "A New Look at Information Literacy at the Massachusetts Institute of Technology." *Presentation at the ASEE Conference*, Chicago, IL, 1997.

    d. In one sentence and in your own words (abstraction of the need statement), define the problem at hand.

    e. How are you going to go about getting your product to customers (e.g., development cost, time, manufacturing, production investment, etc.)?

Establishing who your customers are is one of the most important initial steps that a designer needs to take. As mentioned in Section 1.4, one of the vital concepts to grasp is that customers are not only the end users. Customers of a product are everyone who will deal with the product at some stage during its lifetime. This includes the person who will manufacture the product, the person who will sell the product, the person who will service the product, the person who will maintain the product during its lifetime in operation, etc. Section 1.4 provided an example of possible customers of an airplane. Consider another example: Discuss with your colleagues who the possible customers of a golf cart are. Here are a few ideas to start you off.

- The golf player
- The golf country club (Institution)
- The transportation company that will transport the cart
- The golf club (Equipment) manufacturers for storage of their clubs in the cart

Once all possible customers have been identified, their needs should be considered, and more often than not, their needs can conflict with each other. It is the responsibility of the designer to recognize all of these needs in a prioritized manner and later arrive at a feasible solution that is a an optimal combination of all these 'desires'. One good way to identify the needs in a prioritized manner is to conduct a market survey. There are a number ways in which this can be carried out.

1. Focus group meetings
2. Telephone interviews
3. One-on-one interviews
4. Questionnaires

Each method cited has its advantages and disadvantages. In a focus group meeting, a group of 6 to 12 potential 'customers' meet and discuss their needs and other aspects of the product. If the product already exists, the discussion usually focuses on a 'satisfaction' based feedback in terms of what they liked, what they disliked, and what they would like to see improved. However, for a new product, the discussion usually focuses on their wishes and desires in a particular market segment, what they would like to see introduced to improve their lives, or what current problems exist in the similar products on the market. It is important to ensure that any potential solutions are filtered out at this stage and converted into a neutral need. However, this method is an expensive process, and the sample size is relatively small. It is however a good starting point and is frequently used as a precursor to sending out a larger survey in the form of questionnaires.

Telephone and one-on-one interviews can eliminate some of the ambiguities that arise for questionnaires. However, they are very expensive to run and also have a potential disadvantage of the interviewer 'leading' the interview and causing bias. For example, a question can be asked: "Would you really walk a long distance in the cold, rainy weather, in the middle of rush hour to get to your office early in the morning, or would you prefer taking the cheap, fast, and comfortable public transport?" An unbiased question could be "What is your preferred mode of transport to your office in the morning?"

The questionnaire format is one of the most popular survey methods, as it involves taking the opinion of a large number of people (sample) at a relatively low cost. It is important to construct a questionnaire carefully in order to provide meaningful, useful, and unbiased feedback. Here are some points to follow when creating a questionnaire:

- Develop a standard set of questions. The main goal of a questionnaire is to ascertain potential needs, problems, likes, and dislikes. It is useful at this stage to also identify which (if any) market segment would be most interested in the product as well as to gain an estimate of how much they would be willing to spend.
- Ensure that the questionnaire is easy to read and complete. Use simple language and simple formatting. Try to keep the writing to a minimum, and offer multiple choice questions or yes/no answers where possible. Leave an opportunity for writing for those who wish to do so.
- Identify the demographic you want to target. Mailing lists can be purchased from market research companies.
- Test the questionnaire initially on a pilot sample (friends, family, or small group of people) before sending it out to the entire sample. This is an opportunity to iron out any ambiguous questions and to observe whether or not you are obtaining the desired information.
- Introduce only one issue per question.
- Similar to interviews, you do not want to give your questions a bias. Ensure all questions are unbiased.
- Avoid negative questions, which cause confusion. For example, a question such as "Do you not like to travel in the morning" may result in the answer "No, I do not like to travel in the morning". Reading this carefully reveals a double-negative answer which means "I do like to travel in the morning."
- Ask a few conflicting questions and compare the answers to ensure that the person who has completed the questionnaire actually read the questions. For example ask "Do you ALWAYS switch off the electricity from the mains?" Later on ask "Do you forget to switch off the electricity from the mains?" If the person completing the questionnaire replied the same yes or no to both questions, then this particular feedback is not reliable.

3.4.2 Develop a Strategy

It is important to set a plan for the search process. Looking through every reference book in a library (merely hoping to find the right piece of information) will inevitably lead to an inefficient waste of time. It is helpful first to identify what pieces of information may be needed, and then to select where to begin the search.

Identify Keywords It is important to have some relevant terms with which to begin the search.

Write a Plan It is important to realize that the search process is not linear. For example, while searching the business literature for information on industry trends, you may come across the text of an interview that directly identifies some customer needs for the product under investigation. This information can be very useful if properly contextualized. Thus, it is important to know the framework within which you are working.

3.4.3 Organize and Check the Information Gathered

The following list can be used in two ways: as a planning tool for information collection or as a checklist for locating relevant information.

a. Products
   i. Product names
   ii. Patents
   iii. Pricing
   iv. Parts breakdown
   v. Product features
   vi. Development time
b. Companies
   i. Major players
   ii. Company financials for major players
      • Annual reports—Yearly record of a publicly held company's financial condition. Information such as the company's balance sheet, income, and cash flow statements are included.
      • 10K reports—This is a more detailed version of the annual report and is the official annual business and financial report filed by public companies with the Securities and Exchange Commission. The report contains detailed financial information, a business summary, a list of properties, subsidiaries, legal proceedings, etc.
c. Industry
   i. Trends
   ii. Labor costs
   iii. Market-size industry facts—Pieces of information gained from various sources that help to clarify anything about the industry
d. Market information
   i. Market reports
   ii. Market share of major companies in industry
   iii. Target markets of major competitors
   iv. Demographics
      • Age
      • Geographic location
      • Gender
      • Political/social/cultural factors
e. Consumer trends

3.5 RELEVANT INFORMATION RESOURCES

There are literally hundreds of resources in which information may be located, but several stand out due to their superiority of scope, quality, and overall usefulness. Resources can be divided into four categories.

1. Product information
2. Industry information
3. Company information
4. Market information

Most journal databases will provide information relevant to all of the areas listed in the previous section, and this should be kept in mind when searching and analyzing. This will also help eliminate the repetition of resources in all three areas.

3.5.1 Product Information

Patents grant the patent holder exclusive rights to a new idea or invention. In Chapter 1, Figure 1.5 shows a patent for a machine that makes paperclips, and Figures 1.6 through 1.8 show three different patents of paperclip (1934, 1991, and 1994, respectively). There are two main types of patents: utility patents and design patents. Utility patents deal with how the idea works for a specific function. Design patents only cover the look or form of the idea. Hence, utility patents are very useful, since they cover how the device works not how it looks. Patents can serve as an excellent source of ideas for products and also as a place to protect your ideas should you come up with a novel one. It is important to distinguish that although a patent does prevent others from making and selling your novel idea, a patent does not give you immediate right to make and sell it, since it still has to go through the regular legal and regulatory channels.

Although a good source, it is difficult to identify the specific patent that contains an idea that you are looking for. Since there are about five million utility patents, it is important to use some strategies to hone in on the one that you may be able to use. Use a Web search, such as that provided by the Patent Office. Key word searches as well as patent numbers, inventors, classes, or subclasses are available. Good websites for patent searches include

- http://www.wipo.int/patentscope/en/
- http://www.uspto.gov/patft/index.html

3.5.2 Industry Information

It is essential to know information about the industry your company is considering entering. Who are the major players? What are the current trends? How large (in dollar amounts) are the relevant industries? What materials are used by the relevant industries? The following resources can provide access to information that will answer these questions.

Standard Industrial Classification (SIC) Code SIC codes classify a company's type of business. Many business information sources are organized by the SIC codes. The following publications are guides to these codes.

a. *Standard Industrial Classification Manual*: The official U.S. government manual that provides SIC codes at the two-digit and four-digit levels.

b. *Web version of the SIC code*: This provides access only by SIC category, not subject.

Trade Associations *Encyclopedia of Associations*: Gives address, telephone, and contact information as well as a brief description of trade and industry associations and their publications. Descriptions usually include dates/places for conferences and trade shows.

Industry Overview

a. *Manufacturing U.S.A.: Industry Analyses, Statistics, and Leading Companies*: This annual publication provides statistical data on 459 manufacturing industries, including rations and occupational data. State-level data are included.

b. *Forbes*: Provides an annual report on American industry in the January issue of each year.

c. *Standard and Poor's Industry Surveys*: Provide a textual analysis for 22 broad industry categories which are broken down into more specific subsets.

d. *Encyclopedia of American Industries*: This publication is organized by SIC code with a list of related publications for each described industry.

e. *Moody's Industry Review*: This statistical source contains key financial information and operating data on about 3500 companies. The information is arranged by industry in 137 industry groups. The companies within an industry group may be compared with one another, and they may also be measured against certain averages for that industry. The statistics are updated twice each year.

f. *Predicasts Basebook*: Compiled annually since 1970, but with statistics that go back to 1967, this annual statistical source contains about 29,000 time series which are arranged by a proprietary seven-digit SIC code-based system. The data includes economic indicators and industry statistics. The industry statistics usually include production, consumption, exports/ imports, wholesale prices, plant and equipment expenditures, and wage rates.

g. *Predicasts Forecasts*: This quarterly statistical source provides short- and long-range forecasts for economic indicators as well as industries and products. The data are arranged by a proprietary seven-digit SIC code-based system. Each forecast includes the date and page reference of the source from which the data are taken. The front of each annual accumulation includes composite forecasts, which present historical data for over 500 key series.

Statistics

a. *Annual Survey of Manufacturers*: This survey supplements the Census of Manufacturers in the United States.

b. *American Statistics Index (ASI)*: Indexes and abstracts statistics published by the United States government.

c. *Statistical Reference Index (SRI)*: Indexes and abstracts statistics published by trade and industry organizations and state governments.

3.5.3 Company Information

Company sources provide information on the players within an industry. They provide information on the company financials, products (primary and secondary, such as Pepsi® and Mountain Dew®), brands/trade names, and other "nuggets" that (in context) can help with overall market assessment and analysis.

When searching for company information, it is important to determine whether the company is public or private. Public companies are required to register and file reports with the Securities and Exchange Commission (SEC), while private companies are not. Generally, it is much easier to find information on public companies.

Some companies publish their 10K information on the Web. Examples include

1. Allied Products Corporation
2. American United Global, Inc.

The Web pages contain information extracted from annual and 10K reports of over 12,000 public companies. The data are compiled into a standard format, which allows for sophisticated manipulation of data and customized display formats.

Directories

1. *Hoovers Corporate Directory*: Provides the company's name, street address, phone number, and fax number; the names of the Chief Executive Officer, Chief Financial Officer, and Human Resources director; the most recently available annual sales figures; the percentage change in sales from the previous year; the number of employees; a description of what the company does; and the company's status (private or public).
2. *Directory of Corporate Affiliations*: Lists over 4000 parent companies with divisions, subsidiaries, and affiliates. Provides geographic and SIC indexes.
3. *Duns Million Dollar Disc:* Lists over 160,000 businesses in the United States with net worth over $500,000. Can be used to identify companies by state, industry, or geographic location. Provides thumbnail bios of top executives for public and private companies and basic company information.
4. *Standard & Poor's Register of Corporations, Directors, and Executives, Volume 1; Corporations:* Corporate addresses, telephone numbers, names and titles of officers, SIC codes, products, annual sales, and number of employees for 55,000 public and private companies. Volume 3 provides an index by SIC code.
5. *Corporate Technology Directory:* Directory for high-tech companies with very specific product indexes. Provides basic company information.
6. *Thomas Register of American Manufacturers and Thomas Register Catalog File:* Volumes 1 through 8 list companies arranged by products and services. Volumes 9 and 10 provide an alphabetic list of manufacturers with addresses, subsidiaries, products, and asset estimates; Volume 10 provides an index of brand names. Selected company catalogs are reproduced in Volumes 11 through 16.

7. *Verizon Yellow Pages:* This free service that allows users to look up names, addresses and phone numbers of over 2.1 million businesses.

8. *Ward's Business Directory:* Ward's is a good source for private companies. It provides information on 142,000 companies—90% of which are private. Volumes 1 through 4 provide individual company information. Volume 5 ranks companies by SIC code.

Corporate Reports Moody's manuals also provide detailed information on company history, subsidiaries, business and products, comparative balance sheets, stock and bond descriptions, and more. Geographic indexes classify companies by industry and product. The data in Moody's manuals are obtained from annual and 10K reports.

3.5.4 Market Information

Market information may be found in many places. The search for information should make use of the related industry, company, and product keywords. This is true when using either a paper source or an electronic source. Examples include market share, target market, demographics, and market potential.

Market Research Reports These are ready-made reports that have been commissioned by companies or industrial groups. They are compiled by one of several market research companies. These reports are provided by consulting and advisory firms or by companies and different sectors of industry. The information can be found from firms such as

1. Findex
2. Frost & Sullivan Reports

Market Share and Other Information
1. *Market Share Reporter:* Annual publication providing market share data for companies, products, and industries.
2. *World Market Share Reporter:* Annual publication providing market share information.
3. *Statistical Abstract of the United States:* Statistics on the industrial, social, political, and economic aspects of the United States. A great source for all types of useful statistical information.
4. *Industrial Statistics Yearbook:* Industrial statistics published by the Department of Economic and Social Affairs, which is a statistical office of the United Nations.

Demographic Information
1. *American Demographics:* This journal is searchable on ABI/Inform on First-Search.
2. *State and Metropolitan Area Data Book:* Provides social and economic data for SMSAs, cities, states, and census regions. Prepared by the U.S. Bureau of Census.
3. *Demographics USA, County Edition:* Contains information on purchasing power, consumption rates, economic conditions, and population statistics at http://www.tradedimensions.com

3.6 WEB TOOLS

At this time, the Web does not provide access to all of the resources needed to conduct a product/market research. However, there are some great sites that point you to the resources that are available:

1. Business Researcher's Jumpstation at http://www.brint.com/sites.htm
2. Marketing Resource Directory at http://www.ama.org
3. MIT Resource List at http://me.mit.edu/resources/

Web search engines are increasing in sophistication, as are the Web sites of companies, distributors, industrial association, and retailers. Therefore, it is valuable to spend some time sifting through the information available on the Web.

3.7 CASE STUDY: AUTOMATIC ALUMINUM CAN CRUSHER

We will use this example throughout the remaining chapters of this book. The example was offered as a design project.

3.7.1 Need Statement

Design and build a device/machine that will crush aluminum cans. The device must be fully automatic (i.e., all the operator needs to do is load cans into the device; the device should switch on automatically). The device should automatically crush the can, eject the crushed can, and switch off (unless more cans are loaded). The following guidelines should be adhered to.

- The device must have a continuous can-feeding mechanism.
- Cans should be in good condition when supplied to the device (i.e., not dented, pressed, or slightly twisted).
- The can must be crushed to one-fifth of its original volume.
- The maximum dimensions of the device are not to exceed $20 \times 20 \times 10$ cm.
- Performance will be based on the number of cans crushed in one minute.
- Elementary school children (K and up) must be able to operate the device safely.
- The device must be a stand-alone unit.
- The total cost of the device should not exceed the given budget ($200).

3.7.2 Market Research

Although the need statement is relatively clear, the design team did several interviews with the client, asked questions, and carefully listened to the client's responses in order to determine the goal of the intended device. In parallel, the design team conducted a full market survey to assess similar products as well as to consider all the potential 'stakeholders' or customers.

Here is a short summary of the market analysis.

Potential Customers

- Schools
- Colleges
- Hospitals
- Hotels
- Resorts
- Shopping malls
- Playgrounds and recreational areas
- Apartments, dormitories
- Sports arenas
- Office buildings
- Residential homes

Companies That Have Similar Devices (Selection)

- Edlund Company, Inc. (159 Industrial Parkway, Burlington, Vermont), Mr. R. M. Olson (President)
- Prodeva Inc. (http://prodeva.com)
- Enviro-Care Kruncher Corporation (685 Rupert St., Waterloo, Ontario, N2V1N7, Canada)
- Recycling Equipment Manufacturer (6512 Napa, Spokane, Washington, 99207)
- Kelly Duplex (415 Sliger St., P.O. Box 1266, Springfield, Ohio, 45501)
- Waring Commercial (283 Main St., New Hartford, Connecticut, 06057)
- DLS Enterprises (P.O. Box 1382, Alta Loma, California, 91701)

SIC Code

- Food service industry 3556
- Recycling 3599

Trends

The aluminum industry produces approximately 100 billion cans a year in the USA. This number has been flat for the past 13 years. In 2007, 54 billion of the cans were returned for recycling. According to the Aluminum Association, at a recycling rate of 53.8 percent, the aluminum can is by far the most recycled beverage container in the United States. Although this recycling figure has been rising steadily for the past six years, it actually represents a drop in recycling rates from the previous decade (66.8% in 1997 and 62.1% in 2000), even though the same quantities were produced per year.

3.7.3 Market Information

Growth opportunities for aluminum beverage cans exist throughout the world. Global aluminum can shipments increased 5% annually to 209 billion until the year 2000 and is steadily growing. Seventy-four percent of U.S. beverage sales occurred in convenience stores, drugstores, clubs, mass-merchandise stores, vending machines, and grocery markets. For this product, the initial market is schools. Leon County, Florida public schools, for example, at the high school level, have around 48,846 students.

3.7.4 Patents

Around 56 patents are listed for the past 15 years. A Web-based patent search (Section 4.3.1) will show the details. Here is an example of an old design back in 1981.

US Patent No: 4,436,026, Empty Can Crusher:[2]

*An empty **can crusher** for crushing and flattening empty cans, comprising an inlet, a chute, a stopper device, a pressing device and a forked chute. Empty cans supplied in the crusher are crushed and flattened by the pressing device and are sorted into **aluminium** cans and steel cans by means of a magnet embedded in the pressing device, which fall down into respective receptacles through the forked chute.*

3.8 PROBLEMS

3.8.1   Team Activities

1. How is the direct search different from the indirect search?
2. Define market analysis and discuss how is it different from information gathering.
3. Develop a strategy with which your team will conduct the market analysis task.
4. Write a plan to conduct your market analysis.
5. Estimate the time required for your team to conduct the market analysis assignment.
6. Consider the following need statement: Most houses have vents that open and close manually without any central control. Cities across the country advise the use of vents to help save energy. In most cases, household occupants do not use the entire house at the same time; the tendency is to use certain rooms for a long period of time. For example, the family room and dining room may be used heavily, while the living room and kitchen are used at certain hours of the day. To cool or heat a room, the system must work to cool or heat the entire house. Energy saving can be enhanced if the vents of unused rooms are closed; this will push the hot/cold air to where it is needed most and will reduce the load of the air conditioning system.

   A design team has performed market analysis for this statement but forgot to fill in all of the information. Gather information from available sources and complete the following market analysis.

   I. The SIC codes that are associated with this design are (a) Airflow controllers: air conditioning and refrigeration valve manufacturing (students to provide the codes)

---

[2]*Inventors:* Imamura; Yoshinobu (Nishinomiya, JP), Kamel; Shigeki (Nishinomiya, JP), Yamagata; Tetuo (Kobe, JP), Fujii; Hiroshi (Amagasaki, JP).
Assignee: **Hitachi Kiden Kogyo, Ltd.** (Hyogo, **JP**)
Appl. No.: **06/326,748**
Filed: **December 2, 1981**

(b) Air conditioning units: domestic and industrial manufacturing (provide the codes) (c) Air ducts: sheet metal 5039

II. Materials used in the industry are (a) Sheet metal (b) (provide the codes) (c) (provide the codes) d. (provide the codes)

III. Major associations that deal with air conditioning are as follows: (a) Air Conditioning and Refrigeration Institute (b) (provide the codes) (c) (provide the codes) (d) (provide the codes)

IV. Major companies (a) Trane (b) (provide the codes) (c) (provide the codes)

V. Patent search (a) Sarazen et al., 4,493,456 (b) (provide the codes) (c) (provide the codes)

3.8.2 Individual Activities

1. Perform a Web search and name three industries that change a product at a frequency of at least once each year.

2. How many patents were filled in the last 10 years on money sorters? List a few and identify their differences.

3. List as many Web addresses as you can find that provide patent search. Name some differences among the sites.

4. Interview a librarian in your school and identify research sources that are available for your use.

3.9 Selected Bibliography

http://www.osha.gov/
http://www.ntis.gov/
http://www.census.gov/epcd/www/naicstab.htm
http://vancouver-webpages.com/global-sic/
http://www.thomasregister.com/
http://www.tipcoeurope.com/
http://iml.umkc.edu/ndt/html/lexus.html
http://www.lexis-nexis.com/lncc/
http://www.google.com
http://www.yahoo.com
http://www.briefing.com/
http://www.sme.org/
http://www.asme.org
http://www.manufacturing.net/
http://stats.bls.gov/blshome.htm
http://www.uspto.gov/
http://www.european-patent-office.org/
http://www.nist.gov/
http://www.entrepreneur.com/
http://www.tenonline.org/
http://www.tradedimensions.com
http://www.findex.com
King, HW. J. *The Unwritten Laws of Engineering—Mechanical Engineering.* Vol. 66, No. 7, 1944.

CHAPTER 4

Customer Requirements

Before a detailed product, such as this dashboard, can be designed, a list of needs and goals for such a product should be formulated by the design team. (Volodymyr Kyrylyuk/Shutterstock)

4.1 OBJECTIVES

By the end of this chapter, you should be able to

1. Expand requirements from the needs statement.
2. Prioritize requirements according to importance.
3. Organize requirements into an objective tree.

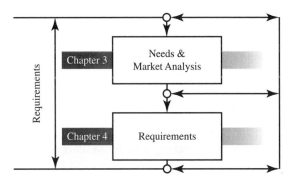

The previous chapter discussed the importance in identifying the needs as defined by all the customers of the product that is to be designed. As needs are essentially a wishlist provided by the customers, the designer has to ultimately translate these needs into a set of specifications that identify how the product will function from a technical standpoint. In order to ensure that these needs are fully addressed as specifications, a middle step is required to aid an accurate transformation. This stage is called the 'Requirements' stage and involves the designer interpreting and prioritizing these needs into product requirements, which essentially identify the objectives of the product.

4.2 IDENTIFYING CUSTOMER REQUIREMENTS

An immediate concern for new readers of this book may well be okay, but what is the difference between a 'need' and a 'requirement'? The answer to this is simple. Needs are a vague set of wishes that customers would like a product to perform for them, such as "Get me from point a to point b as quickly and safely as possible", whereas requirements are the designers detailed breakdown of what the product should do and achieve yet WITHOUT providing solutions. It is essentially an expanded and more organized form of the initial needs. This is the reason why they are still regarded as 'customer requirements.' Some customers may even provide enough detail of their needs that warrants these items to be moved over unchanged to the requirements stage.

So, for example, a need from a customer could be "Something that will hold sufficient quantities of water, have the ability to heat the water efficiently, and have a way in which to pour this water into a mug or cup safely without spilling or burning." At first sight, you may immediately think of an electric kettle, but it is important at this stage not to jump to any conclusions or solutions. The next stage would be to research your market and obtain more information on the customers of this product. This will then enable you to identify the type, frequency, and quantity of usage. Indeed, you may realize, if this will be used in a commercial environment, that the quantity of water that is regarded as sufficient only can be achieved in a large dispenser-type machine. Even if the product was for domestic use, there are many other ways to provide energy to heat up water as well as varying shapes and vessels to hold and pour the water. An extract of our solution-neutral requirements may look like this:

- Hold varied quantities of water
- Heat varied quantities of water
- Boil water fast
- Energy efficient
- Easy to move around
- Safe handling during pouring
- Pour hot water without spilling
- Aesthetically pleasing surface
- Automatic switching off from the energy source or alert user when water is boiling
- etc.

4.3 PRIORITIZING CUSTOMER REQUIREMENTS

Once a list of requirements has been established, the next two steps are to prioritize and organize these requirements so that the designer is aware of the essential requirements as well as the ones that can be compromised due to conflict, cost or other reasons. Conflicts may arise sometimes when a customer wishes for more than one feature that the same product cannot provide; e.g., a portable travel kettle that can hold 10 liters of water at any one time! In this case the designer must identify whether the priority for the customer is the portability or the ability to hold 10 liters of water or if there is a way to compromise and/or bias one requirement over the other.

In order to prioritize requirements, the designer assigns an importance rating for each requirement from 1 to 10, where 10 is the most important and 1 is the least important. Another distinction that is usually made is whether a particular requirement for the product is essential or not. If it is considered essential, then it is classified as a 'Demand' and denoted with a letter 'D'. A 'Demand' is always given the top rating value of '10'. Other non-essential requirements are considered to be 'Wishes' and are denoted with a letter 'W'. The ratings for these 'Wishes' vary from 1 to 10 as described above.

Usually the designer would use feedback from the customers and market research to determine the importance rating from each requirement. As a group team working on a design project, you could rank these wishes in order or conduct a survey in the event that the product you are working on will be beneficial to a group of customers that are easily identified and accessed. Companies desiring to conduct such ordering of wishes use surveys. To distinguish numerically the importance of each wish, a weighing factor is used. There are different methods to allocate weighing factors for these wishes.

a. Absolute measure for each wish individually, where each wish is rated from 1 to 10. Wishes that have the same importance are assigned the same value. The total measure is irrelevant.

b. Relative measure, where a total of 100 is maintained and each wish is given a value according to its importance. The weighing factor in this event could be a percentage figure adding to 100% when adding all values.

In many instances the customer will be more than one entity. (For example, the brake pads for a car would have the drivers, the mechanic, and salespeople as essential customers who may view and evaluate attributes differently.) In such events the attributes will be evaluated in three different columns according to their importance. Thus all three customer categories will have to input the value for the weighing factor as they see fit. Students need to be aware of this strategy; however, in many design cases where students compose the design team and the instructor or an industry representative is the customer, it is reasonable to ask the customer and the design team to grade the wishes.

This importance rating will later be used to quantitatively assess how well the design of the product is progressing using a popular technique called Quality Function Deployment (QFD), which is covered in more detail in Chapter 6.

EXAMPLE 4.1 Wheelchair Retrieval

Design a wheelchair retrieval unit to assist nurses in situations where a patient is taking a walk with his nurse and then in about 30 m feels tired. The nurse must remain with the patient to support him and should be able to use one hand to activate the wheelchair retrieval unit.

Solution

Design the wheelchair retrieval unit with the following requirements:

- Unit dimension within $30 \, \text{cm}^3$
- Reach the patient in 1 minute
- Responsive within 30 m away from the nurse

Attention must be paid to appearance. The finished product should be marketed in two years' time. Manufacturing cost should not exceed $50 each at a production rate of 1000 per month. In the design process, after the designers have created the objective tree and the function tree, it is important that the designers complete a table of specifications. Designers must decide on the level of generality of the specifications. The requirements are shown in Table 4.1.

This example is for a wheelchair retrieval unit that can be operated with one hand and help a nurse who is supporting a patient to retrieve the wheelchair. The initial statement has been considerably expanded. The range of users has also been taken into account, as have safety considerations. The demands (D) and wishes (W) have been distinguished.

TABLE 4.1 Prioritized for a Wheelchair Unit

D or W	Requirements	Importance (1–10)
D	Dimension $30 \times 30 \times 30 \, cm^3$	10
D	Reach patient in 1 min	10
D	Water resistant	10
D	Stable	10
D	Ability to support patient weight	10
D	Ability to provide transportation & seating	10
D	Easy to operate	10
D	Low maintenance	10
D	Durable	10
D	Safe	10
W	Stops quick	7
W	Comfort	6
W	Easy to control	7
W	Low power drain when not in operation	5
W	Minimum amounts of parts	6
W	Lightweight	4
W	Compact/foldable	3
W	Low production cost	7
W	Operate by itself	3
W	Small turning radius	5
W	Maneuver through obstacles efficiently	8

4.4 CASE STUDY: AUTOMATIC ALUMINUM CAN CRUSHER—REQUIREMENTS

Continuing our case study which we started in Section 3.7, we shall be considering the automatic aluminum can crusher. As a reminder, the following was the original need statement:

Design and build a device/machine that will crush aluminum cans. The device must be fully automatic (i.e., all the operator needs to do is load cans into the device; the device should switch on automatically). The device should automatically crush the can, eject the crushed can, and switch off (unless more cans are loaded). The following guidelines should be adhered to:

- The device must have a continuous can-feeding mechanism.
- Cans should be in good condition when supplied to the device (i.e., not dented, pressed, or slightly twisted).
- The can must be crushed to one-fifth of its original volume.
- The maximum dimensions of the device are not to exceed $1.5 \times 1.5 \times 1$ feet.

- Performance will be based on the number of cans crushed in one minute.
- Elementary school children (K and up) must be able to operate the device safely.
- The device must be a stand-alone unit.
- The total cost of the device should not exceed the given budget ($200).

Once the detailed market research was completed (Section 3.7), the engineering team met and decided on whether each requirement was a demand (D) or wish (W) and allocated importance ratings and finally constructed the list of prioritized requirements as given in Table 4.2.

TABLE 4.2 Aluminum Can Crusher Requirements

List of requirements in no particular order

Requirement	D/W	Rating	Requirement	D/W	Rating
Pleasing to eye (stylish and fashionable)	W	5	Stand-alone unit	D	10
Internal parts totally enclosed: safe from ambient environment	D	10	Large capacity feeder and refuse container	W	4
Blends with surrounding	W	5	Cans automatically removed	D	10
Dimensions $1 \times 1.5 \times 1.5$	D	10	Low peripheral force	W	8
Inconspicuous	W	5	Easy to start	W	9
Many colors available	W	1	Shock absorption	W	8
Built from polymer	W	2	Easy cleaning	W	3
Housing constructed from molded polymer	W	2	Durable refuse container	W	4
Paintable surface	W	3	Can counter	W	2
Ability to reset after kill switch has been used	W	9	Receiving containers on casters	W	3
Plexiglass window to view operation	W	1	Sealed bearings	W	7
Low noise	W	7	Ability to mount to various surfaces	D	10
Machine rendered inoperable when opened	W	10	Compact size	D	10
Operable by elementary students	D	10	Ability to crush various sizes of containers	W	3
Total Cost < $200	D	10	Operator-free operation	W	6
Automatic kill switch and reset button within easy reach	W	10	Enough force to crush cans	D	10
Wiring kept away from moving parts	D	10	Starts immediately	W	7
Crushing mechanism inaccessible from feeder & dispenser	W	10	High efficiency engine	W	9
Yellow light to indicate improper use of machine	W	3	Light weight	W	3
Internal parts safe from liquid damage	W	8	Low loading height	W	7
Stops easily and immediately	W	9	Flip open lid	W	1
Little heat produced	W	6	Solvents unable to hurt finish	W	6
No sharp corners	W	8	Limit # of light tolerances	W	7
Ability to stop in mid-operation	W	10	Many cans crushed per minute	W	8
Green light to indicate it is okay to load	W	3	Reduce volume by 80%	D	10
Extra wiring insulation	W	6	Can glass and plastic crushing	W	5
Operation and safety stickers	W	9	Runs on standard 110V outlet	W	9
Red light to indicate crushing mechanism is in operation	W	3	Long running capability	W	9
No flying debris	W	10	Consolidate mechanical functions	W	5
No exhaust	W	8	Portable	W	4
Continuous	D	10	Easily assessible interior	W	7
Easy access to clear jams	W	6	Container to hold refuse liquid	W	2
Easy to maintain and disassemble	W	3	No/little service required	W	3
Low vibration	W	8	Utilizes ground to stablize	W	5
Variable length retractable cord	W	3	Small force required to depress switches	W	9
Stops easily and immediately	W	8	Stand-by mode	W	3
Utilizes gravity	W	2	Weather proof	W	3
High material strength	W	9	Less than five assembly steps	W	5
Retails for < $50	W	3	Large storage of crushed cans	W	3

4.5 ORGANIZING CUSTOMER REQUIREMENTS—OBJECTIVE TREE

The next important step at this stage of the design is organizing the customer requirements, which helps clarify the objectives of the design. A popular method in organizing the customer requirements is by developing an objective tree. An objective tree allows a clear and concise method in representing the requirements of the project to be carried out. It will also help to minimize any confusion between the customers and the design team as both should agree on the finalized objective tree, which illustrates, in diagrammatic form, the ways in which different objectives are related to each other. The objective tree method procedure was summarized by Cross.

1. Prepare a list of design objectives. These take the form of design briefs, prepared from questions to the client and from discussion with the design team. Remember to ask as many questions as you possibly can to enable you to better understand what exactly the customer needs. Remember, a vague statement is equal to a vague understanding of the need, which may lead you to develop a product that does not match the customer needs. This is the most important step in developing the tree. Further discussion among the design team of what you would like to have in the product is important. Remember, there is no limitation for what you can put in the product at this time.
2. Order the list into sets of higher-level and lower-level objectives. The expanded list of objectives and sub-objectives is grouped roughly into hierarchical levels.
3. Draw a diagrammatic tree of objectives, showing hierarchical relationships and interconnections. The branches in the tree represent relationships, which suggest means of achieving objectives.

EXAMPLE 4.2 Water Purifier

After the residents of Gotham City complained to their mayor about the city's water quality, he ordered the health unit to investigate the complaint. The health unit's recommendation was based on the chemical analysis done by the top chemist at the Department of Public Health. The task is now given to the city engineers to design water purifiers. Your task is to help them build an objective tree.

Solution

Step 1. Prepare a list of design objectives.
   a. Cost effectiveness
   b. Safety

    c. Can detect chemical imbalance
    d. Fewer repairs
    e. Easy to repair when needed
    f. Long lasting
    g. Affordable
    h. Low damage
    i. Low or no contamination
    j. Takes up least possible space
    k. Safe for humans
    l. Safe for environment
    m. Gets the job done
    n. Can correct problems in least time
    o. Low maintenance
    p. Cleans high volume of water
    q. Efficient
    r. . . . . (complete per project objectives)
    s. . . . .

Step 2. Order the list into sets.

Safety	Cost Effectiveness	Efficiency
Safe for humans	Few repairs	Can detect chemicals
Safe for environment	Easy to repair	Long lasting
	Affordable	Low damage
	Takes least possible space	Gets job done
	Low maintenance	Corrects problems in minimal time

Step 3. Draw an objective tree (Figure 4.1).

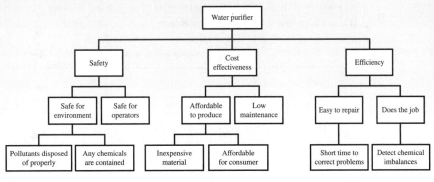

Figure 4.1   Objective tree for water purifier.

EXAMPLE 4.3 Automatic Coffee Maker

Build an objective tree for an automatic coffee maker.

Solution

Step 1. Prepare a list of design objectives.
- a. Safety
- b. Efficiency
- c. Quality
- d. Convenience
- e. Easy to use
- f. Fast
- g. Makes good coffee
- h. Doesn't burn user
- i. Good mixture
- j. Right temperature
- k. Splash proof
- l. Cheap to consumer
- m. Volume of coffee
- n. Automatic
- o. Timer
- p. Energy saver
- q. Temperature control
- r. Easy to clean
- s. . . . .(complete per project objectives)
- t. . . . .

Step 2. Order the list into sets.

Safety	Quality	Convenience	Economical
To user	Coffee tastes good	Easy to clean	Affordable
Doesn't burn user	Right temperature	Automatic	Volume of coffee
Splash proof	Good mixture	Timer	Energy saver
	Temperature control		Fast

Step 3. Draw an objective tree. (See Figure 4.2.)

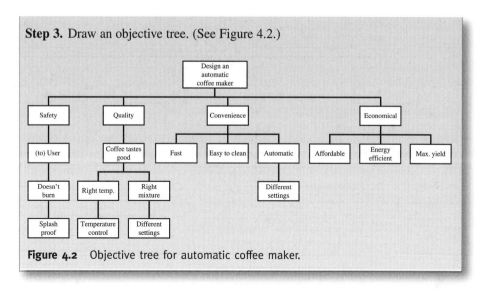

Figure 4.2   Objective tree for automatic coffee maker.

4.6 CASE STUDY: AUTOMATIC ALUMINUM CAN CRUSHER – OBJECTIVE TREE

Continuing our case study which we started in Section 3.7. we shall be considering the automatic aluminum can crusher. The Need Statement and Market Research was carried out in Section 3.7, and the list of prioritized requirements was generated in Section 4.3.

The next stage for the design team was to try to make sense of these requirements in an organized fashion and hence distribute the list of requirements into an objective tree. The objective tree shown in Figures 4.3 through 4.7, clearly states the goals and defines the directions of the device design.

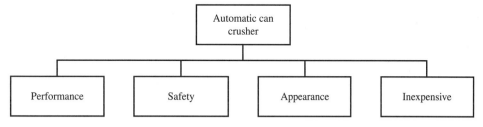

Figure 4.3   Main heading.

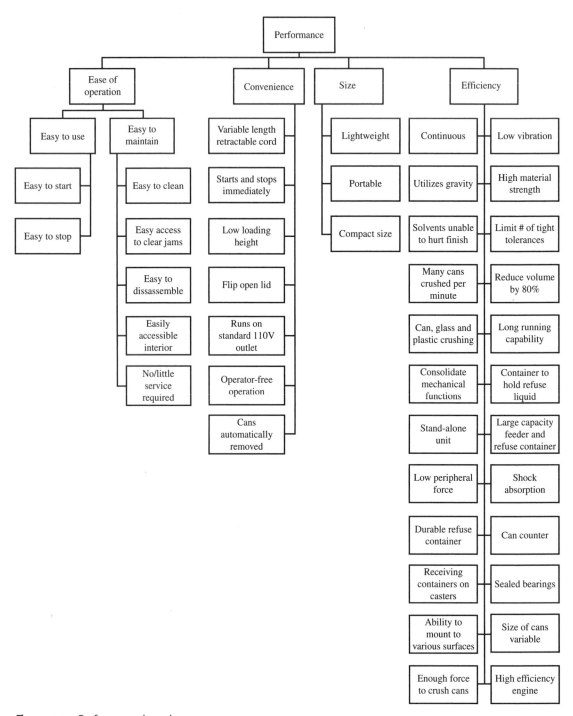

Figure 4.4 Performance branch.

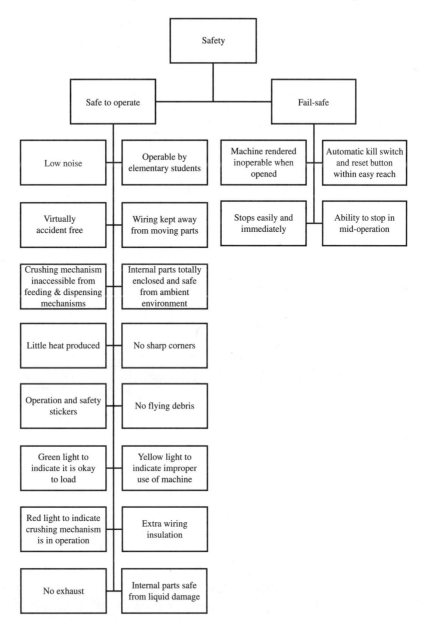

Figure 4.5   Safety branch.

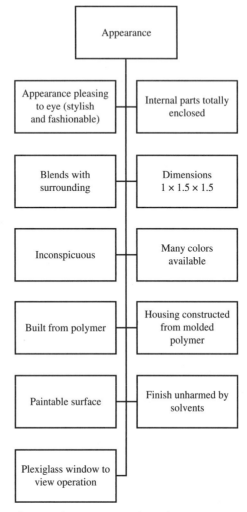

Figure 4.6   Appearance branch.

LAB 7: Kano Model Customer Needs Assessment

This lab deals with Kano model-based customer needs assessment. The Kano model defines customer needs based on customer satisfaction (see Figure L7.1).

As shown in Figure L7.1, there are five types of customer needs: (1) *must-be*, (2) *one dimensional*, (3) *attractive*, (4) *indifferent*, and (5) *reverse*. A customer need in terms of a product attribute is considered *must-be* if its absence produces absolute dissatisfaction and its presence does not increase satisfaction. An attribute is considered *one dimensional* if its fulfillment increases the satisfaction and vice versa. An attribute is considered *attractive* if it leads to a greater satisfaction—it is not expected to be in the product. An attribute is considered *indifferent* if its presence or absence does not contribute

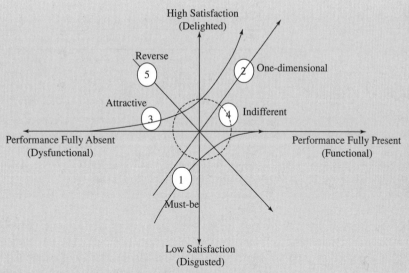

Figure L7.1   Definition of customer needs.

to the satisfaction. An attribute is considered *reverse* if its presence causes dissatisfaction and vice versa. Thus, to develop a customer-focused product, it is important to do the following:

(a)   *keep the must-be attributes,*

(b)   *add a good number of one-dimensional and attractive attributes,*

(c)   *avoid indifferent attributes as much as possible*, and

(d)   *avoid the reverse attributes.*

To know the preference of customers, the Kano model provides a questionnaire, as shown in Table L7.1. A customer selects one answer (Like, Must-be, Neutral, Live-with, or Dislike) from the *functional* side and the other from the *dysfunctional* side to assert his/her preference. Moreover, the Kano model provides a definition of consistency from the customer answers, as shown in Table L7.2. For example, if a customer selected 'Like' from the functional side and 'Neutral' from the dysfunctional side, it means that the attribute is 'attractive' to customer needs.

TABLE L7.1 Example Questionnaire

Product: **Room Heater** *Attribute:* **Steamer**	*Functional:* Room Heater with a Steamer	*Dysfunctional:* Room Heater without a Steamer
	Like	Like
	Must-be	Must-be
	Neutral	Neutral
	Live-with	Live-with
	Dislike	Dislike

TABLE L7.2 Customer Needs Evaluation

| | Dysfunctional | | | | |
	Like	Must-be	Neutral	Live-with	Dislike
Functional	Q	A	A	A	O
Must-be	R	I	I	I	M
Neutral	R	I	I	I	M
Live-with	R	I	I	I	M
Dislike	R	R	R	R	Q

Attractive (A), Indifferent (I), Must-be (M), One-dimensional (O), Questionable (Q), and Reverse (R)

Table L7.3 shows customer answers for three attributes of a mobile phone: 'keypad-display-same-side', 'keypad-small-display-large', and 'keypad-large-display-small.' The intention is to identify the customer preference regarding the relative size of a keypad and display of a mobile phone.

Figure L7.2 shows how the status of 'keypad-display-same-size' is determined from the answers of fifteen customers using Kano evaluation in Table L7.2. As seen Figure L7.2, this customer's attribute is 'Indifferent' (i.e., does not help much to increase his/her satisfaction). There are a relatively large number of customers that consider it to be 'Reverse' (i.e., does not want keypad and display to be equal in size). Therefore, the product developer should avoid this attribute and find solutions from the other two.

GROUP TASK Find out the status of the two other attributes in Table L7.3, and determine what the relative size of the keypad and display of mobile phone should be according to your overall evaluation.

TABLE L7.3 Customer Answers for Three Attributes of a Mobile Phone

Customer	Keypad-display-same-size Functional	Dysfunctional	Keypad-small-display-large Functional	Dysfunctional	Keypad-large-display-small Functional	Dysfunctional
1	Dislike	Must-be	Must-be	Dislike	Dislike	Must-be
2	Live-with	Neutral	Like	Dislike	Dislike	Must-be
3	Dislike	Like	Like	Dislike	Dislike	Like
4	Live-with	Neutral	Must-be	Dislike	Dislike	Must-be
5	Neutral	Live-with	Like	Neutral	Dislike	Must-be
6	Must-be	Live-with	Like	Neutral	Dislike	Must-be
7	Dislike	Like	Like	Dislike	Dislike	Like
8	Neutral	Neutral	Must-be	Dislike	Dislike	Must-be
9	Like	Dislike	Must-be	Live-with	Dislike	Must-be
10	Neutral	Neutral	Must-be	Dislike	Dislike	Like
11	Dislike	Must-be	Must-be	Dislike	Dislike	Must-be
12	Must-be	Dislike	Must-be	Live-with	Dislike	Must-be
13	Neutral	Neutral	Dislike	Like	Dislike	Like
14	Like	Live-with	Like	Live-with	Live-with	Like
15	Like	Neutral	Like	Live-with	Live-with	Neutral

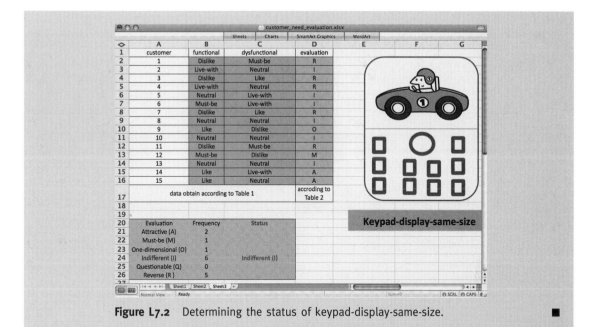

Figure L7.2 Determining the status of keypad-display-same-size.

4.7 PROBLEMS

4.7.1 Team Activities

1. Use the statement "Connect two bodies of water with no gradient." Develop a series of questions to better understand the need.

2. Use the following statement: "Ice is forming on the roof with density gradient." Develop a series of questions to better understand the need.

3. In this activity each group, as shown in the following list, asks as many questions as they can to clarify the given design statement:
   (a) Groups 1 and 9: Design a coffee maker.
   (b) Groups 2 and 10: Design a safe ladder.
   (c) Groups 3 and 11: Design a safe chair.
   (d) Groups 4 and 12: Design a safe lawn mower.

4. Draw an objective tree for an automatic tea maker.

5. Draw an objective tree for the following:
   If you drive in the state of Florida, you may notice some traffic congestion due to the trees being trimmed on the state roads and freeways. Assume that the Florida Department of Transportation is the client. Usually, branches 6 inches or less in diameter are trimmed. The allowable horizontal distance from the edge of the road to the branches is 6 ft. The material removed from the trees must be collected and removed from the roadside. To reduce the cost of trimming, a maximum of only two workers can be assigned for each machine. The overall cost, which includes equipment, labor, etc., needs to be reduced by at least 25% from present cost. The

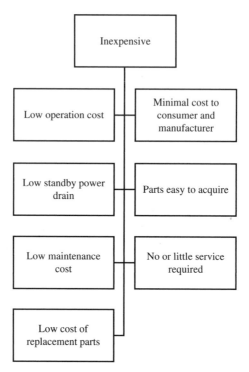

Figure 4.7   Inexpensive branch.

state claims that the demand for your machines will follow the price reduction (i.e., if you are able to reduce the cost by 40%, the demand will increase by 40%). Allowable working hours depend on the daylight and weather conditions.

6. Draw an objective tree for the following statement:
   We have a mountain-sized pile of wood chips that we want to process into home-use fire logs. We are located in Tallahassee, Florida. We can have a continuous supply of wood chips throughout the year. We should be able to produce 50 fire logs per minute. The shareholders require us to have a large profit margin, and our prices should be lower than our competition for the same size logs. The logs should produce enough heat to keep our customers faithful, and they should have environmentally clean exhaust.

7. As technology advances and new materials are synthesized, it seems that there are certain products that receive only minimal, if any, modernization attempts. Among these are the shopping carts that are used in grocery stores. There is a tendency to save parking spaces by not designating a return cart area. Leaving carts in the parking lots may lead to serious accidents and car damage. Furthermore, many customers do not fill their carts when shopping; however, they do not like to carry baskets. Other customers like to sort products as they shop. Develop an objective tree to clarify the need statement and prioritize the objectives laid out in the problem statement. You may add features that may not be listed clearly in the problem statement but will give an additional advantage to the proposed design.

8. Most houses have vents that open and close manually without any central control. Cities across the United States advise the use of such vents in an effort to save energy. In most cases, household occupants do not use the entire house at the same time; the tendency is to use certain rooms for long periods of time. For example, the family room and dining room may be used heavily, while the living room and kitchen are used at certain hours of the day. To cool or heat a room, the vent system must work to cool or heat the entire house. Energy saving can be enhanced if the vents of unused rooms are closed; this will push the hot/cold air to where it is needed most and reduce the load of the conditioning system. Develop an objective tree to clarify the need statement and prioritize the objectives laid out in the problem statement. You may add features that may not be listed clearly in the problem statement but will give an additional advantage to the proposed design.

4.8   Selected Bibliography

CROSS, N. *Engineering Design Methods: Strategies for Product Design.* New York: Wiley, 1994.

DYM, C. L. *Engineering Design: A Synthesis of Views.* Cambridge, UK: Cambridge University Press, 1994.

HENSEL, E. "A Multi-Faceted Design Process for Multi-Disciplinary Capstone Design Projects." *Proceedings of the 2001 American Society for Engineering Education Annual Conference and Exposition*, Albuquerque, NM, 2001.

KARUPPOOR, S. S., BURGER, C. P., and CHONA, R. "A Way of Doing Design." *Proceedings of the 2001 American Society for Engineering Education Annual Conference and Exposition.* Albuquerque, NM, 2001.

PAHL, G., and BEITZ, W. *Engineering Design: A Systematic Approach.* New York: Springer-Verlag, 1996.

SUH, N. P. *The Principles of Design.* New York: Oxford University Press, 1990.

CHAPTER •5

Establishing Functional Structure

It is important to determine a product's functional structure before completing the design. Engineers will determine the function of each of this drill's components in order to determine all possible solutions in its final design. (Kolosigor/Shutterstock)

5.1 OBJECTIVES

By the end of this chapter, you should be able to

1. Discuss the essentials of function analysis.
2. Create function structures based on need statements.
3. Discuss the importance of function analysis.

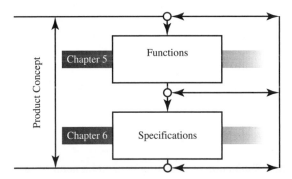

In previous chapters, you have worked your way to identifying, clarifying, and prioritizing the objectives for a design project (requirements phase). Once this is done, the next stage is to begin developing the product concept. This will be carried out in two main sections: functions (this chapter) and specifications (Chapter 6).

At present, there are several design strategies that are utilized by design teams. In many cases, engineering designers develop their own methods, usually based on their experience and current practices in their working environment. However, when engineers are faced with a novel problem outside their immediate experience, they search furiously for an organized systematic procedure. Several theories and methodologies have been proposed for design. In one theory, it is argued that designers are born with a creative mind that is capable of producing inspired designs. In this theory, design cannot be taught or learned; it is a gift bestowed on unique and privileged individuals. Fortunately, this theory is not widely accepted.

In another design philosophy, once a need is identified, a concept is directly developed by brainstorming. However, this philosophy often does not lead to quality designs, and it does not foster the use of the latest technologies. In this methodology, designers have a tendency to take their first idea and keep refining it toward a product design. In this book, we utilize a more scientific view of design: a systematic approach in logical sequence. This approach is in agreement with what is known as the "industry best practice."

In the process you have followed thus far, you are still defining the problem utilizing different tools. You have developed an objective tree; you have searched the market; and you need to start developing an idea on how the product could function. This essentially involves converting your customer requirements into 'functions.'

5.2 FUNCTIONS

The overall function of a product is the relationship between its inputs and output. The function of the product can be further broken down to subfunctions that identify purposive actions that the product is meant to perform. Whereas requirements, as set by the customer, are 'wish-lists' that describe what the product should do, functions are solution-neutral engineering actions that the product will perform. This stage is important in several ways. The first is that this stage signifies converting wishes into engineering terminology which is more relevant to the design team. The second is that it is important to realize that functions remain solution-neutral, and hence, it is still a means to further ascertain the problem. It may seem that a lot of time is spent at identifying and refining the problem. At first sight; this may seem like an inefficient process. This is so much so that some inexperienced designers become impatient and decide to short-cut or even skip this stage and try to suggest different concepts or solutions immediately. The thing to remember here is that it is unlikely that the best solution can be provided to something where the best definition of the problem is not available.

Functions should consider 'what' the product does (the problem) and not 'how' it does it (solution). A function involves the following two components.

- An action verb
- A noun representing the object on which the action verb takes place

The combination of an action verb and a noun are used to describe any given function.

5.3 FUNCTION DECOMPOSITION AND STRUCTURE

Functions should be broken down as finely as possible. This process is known as *functional decomposition* and is represented as the *functional structure* of the product. A functional structure consists of the following:

- A boundary box (with inputs and outputs)
- An overall function
- Function tree
- Known flow of materials, energy, and information

Note that although materials are mentioned here, these are not necessarily the detailed material properties of the product as you still should have no idea on the concept let alone the details of the product. Materials here refer to the 'items' that are known and are needed to perform the actions. Therefore for a coffee machine, it is a given that coffee, water, and a container to hold the beverage will be needed. It is not of concern now what material the coffee machine is made out of.

5.3.1 Bounding Box and Overall Function Diagram

The simplest form of a functional structure of a product is represented as the overall function diagram. This is commonly referred to as a "black box" and presents just the overall

Inputs   Outputs

BLACK BOX

Black Box System Model

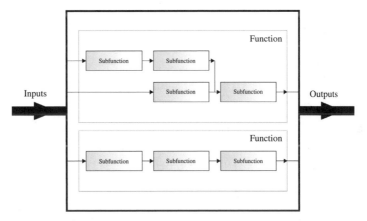

Transparent Box Model

Figure 5.1   The box diagram.

function of the product along with inputs and outputs to the system, which can include flow of energy, materials, and information from/to the system surroundings. The subfunctions will not appear in this diagram. This form of presentation expresses the relationship between inputs and outputs regardless of the solution within the box. A simple box diagram is illustrated in Figure 5.1. Figure 5.1a represents the concept of the functional structure as a black box where a set of inputs are performed by an overall function (consisting of many subfunctions—not depicted) and a set of outputs. Figure 5.1b shows a transparent version of the box model depicting the subfunctions of the product. The subfunctions here are known as the *function tree* of the product and essentially identify the actions and sequences needed to perform the product's main (overall) function.

Consider the objective tree of the automatic coffee machine that was described in Chapter 3 (Figure 3.2). Let us begin the process of establishing a functional structure for this machine. Let us assume that market research identified the need for an automatic coffee machine that enables grinding fresh coffee beans as well as mixing milk and sugar according to the user's preference. The inputs would be water, coffee beans, sugar, milk, the container to hold the hot beverage, as well as a power source to heat the water. Notice that the term 'power source' is mentioned instead of 'electricity'. Specifying 'electricity' would be providing a solution, and our goal at this stage is to remain solution-neutral. It also would be limiting our options as other power sources could be solar power, gas, or even nuclear energy in the future! Why restrict ourselves now? The advantage of remaining solution-neutral at this stage is that if new technologies become available in the future,

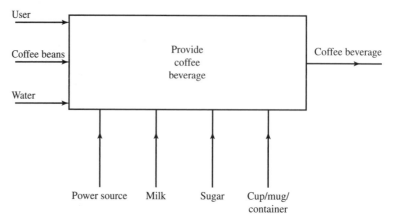

Figure 5.2 Overall function diagram for an automatic coffee machine.

they can be substituted easily when they become available and feasible while still maintaining the same functions of the product to achieve the desired results.

The output of the coffee machine will be the coffee beverage. The main function of the machine is to provide the coffee beverage.

Figure 5.2 shows the overall function diagram of the automatic coffee machine, which includes the inputs, outputs, overall function, and the black box.

5.3.2 Function Tree

A complex, overall function describing the problem statement can be broken into a number of functions and subfunctions. Many of the subfunctions can be divided into sub-subfunctions. The division depends on the type of design system being considered. The items within the black box (the subfunctions) are organized in a systematic and logical sequence to enable the execution of the individual actions that are needed to perform this main function. This is known as the function tree. The main function of the automatic coffee machine is to provide the coffee beverage. The function tree of the automatic coffee machine is shown in Figure 5.3.

Several guiding points to follow when attempting to create a function tree are the following.

- The flow should be in a logical or temporal order.
- Identify redundant functions and combine them.
- Functions not within the system should be eliminated.

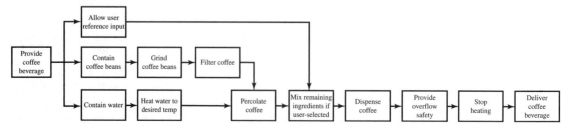

Figure 5.3 Function tree for the automatic coffee machine.

Specifically, there are flows associated with materials, energy, and information. Some functions associated with the flow of materials are as follows.

1. Material can be designed to alter its position or shape. Usual action verbs are *lift, position, hold, support, move, translate, rotate,* and *guide.*
2. Material can be divided into two or more bodies. The action verbs are *disassemble* and *separate.*
3. Material assembly. Usual terms are *mix, attach,* and *position relative to.*

Functions associated with the flow of information are usually in the form of mechanical or electrical signals or software.

Functions concerned with the flow of energy in electromagnetic systems are mechanical, electrical, fluid, and thermal. Energy of this type can be supplied, stored, transformed, or dissipated.

5.3.3 Function Structure

The previous section demonstrated the functional decomposition of the automatic coffee machine. The culmination of this process is the functional structure and is shown in Figure 5.4. This is the combination of the box, inputs and outputs, flow of materials, and the function tree. The primary purpose of function structures is to facilitate the discovery of solutions. Task-specific functions, provided in the following list, can be utilized in generating a functional analysis.

1. Conversion of energy
   a. Changing energy (e.g., electrical to mechanical energy)
   b. Varying energy components (e.g., increasing speed or torque)
   c. Storing energy (e.g., storing potential or kinetic energy)
   d. Connecting energy with information (e.g., switch to start a motor)

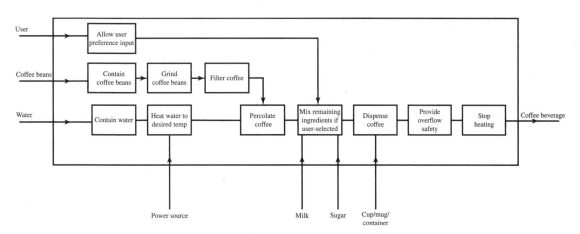

Figure 5.4 Functional structure for the automatic coffee machine.

2. Conversion of materials
   a. Changing matter (e.g., melting [solid to liquid])
   b. Connecting matter with energy (e.g., moving parts)
   c. Rearranging materials (e.g., mixing or separation)
   d. Storing materials (e.g., storing material in a silo)
3. Conversion of information
   a. Changing signals (e.g., mechanical to electrical)
   b. Connecting information with energy (e.g., amplifying signals)
   c. Connecting information with matter (e.g., marking metals)
   d. Storing signals (e.g., data banks)

In original designs, neither the individual subfunctions nor their relationships are clearly known. Hence, in such a case, a strong effort to search for and establish an optimum function structure is required. However, in adaptive designs, general function structure and the functional relationships between various components and systems usually are well established. The lower levels of functions carry smaller levels of complexity and should be simple to design. Function structures are of great importance in the development of modular designs. An important advantage of a function structure is that each of the components and systems can be examined individually and modified if needed. Function structures are also very useful when designing and using design catalogs, because for every function, a task-specific component or system can be identified. The whole process of functional decomposition supports the divergent–convergent design philosophy. This concept requires that a design problem be expanded into many solutions before it is narrowed to one final solution. All design problems should be decomposed into the design of functionally independent subsystems.

Solution principles or designs based on conventional methods do not provide optimum designs when new technologies and discoveries are to be used. The crux of a problem should be expressed as a list of the required functions and essential constraints.

Let's refer to another one of our examples—the automatic can crusher. The design team reviewed the objective tree and their market analysis findings before they started structuring a function tree. They kept their minds open to all possible methods for accomplishing the stated goals. The function structure is shown in the Figure 5.5.

As you observe from the figure, once again the system is enclosed within a boundary (the box) that allows a flow of energy, input of cans, and output of crushed cans to enter through this boundary. There are a series of functions connected from the input side to the output side that compose the whole system. Loading of cans is utilized to input the cans

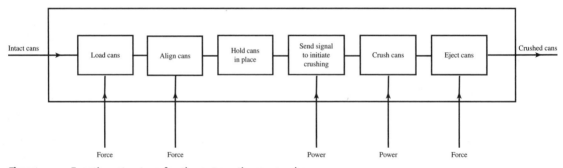

Figure 5.5 Function structure for the automatic can crusher.

into the system. The cans will then be aligned so as to position them inside the device for crushing. A method to hold the cans in place is needed during the crushing. Power is then needed to send a signal to the crushing device. Then, after crushing is performed, the cans must be removed or ejected from the system. Once again, note that no solution or indication of how to solve the problem is presented at this stage.

You should ask at this point, "Is this the only set of functions that could perform the essentials of crushing cans?" The answer is no. However, it should be emphasized that the systematic design process is an iterative one. Further proposing a system to crush cans in terms of functional needs will satisfy the customer need statement, which is redefined through the objective tree. There could be some functions and requirements in the objective tree that no action function could satisfy, such as the objective of having a safe device. This requirement, however, will be further clarified in later functions. At this stage, the design team should revisit the objective tree and revise the tree if additional functions are found to be necessary. Thus, the iterative process is demonstrated.

Design teams should be alerted not to generate an idea and then fit the function analysis to that one idea. At this stage, you should have no constraints on what could be implemented in the design and what could not. Students who tend to bend the function analysis to fit an agreed-on and hidden idea tend to spend more time refining the one idea to fit the procedures and steps needed in the future stages of the design process. Creating and identifying functions is an essential step in the design process; the later design steps depend on the function analysis step. Thus, revisiting this step and reiterating is highly recommended.

5.4   DETAILED PROCEDURE TO ESTABLISH FUNCTIONAL STRUCTURES

This section formalizes what has been discussed in the previous sections and thus provides a step-by-step procedure as outlined by Cross in carrying out functional decomposition of a design and representing it as a functional structure:

Step 1. *Express the overall function for the design in terms of conversion of inputs and outputs:* The key is to determine what needs to be achieved by the new design and how it is to be achieved. This can be accomplished by representing the product or device (to be designed) simply as a black box (Figure 5.5), which converts given inputs to desired outputs. The black box specifies all the functions required to convert the input into output. It is important not to place any restrictions on the function. Such restrictions limit the solution space or system boundary. For example, the function "push from A to B" restricts the solutions from ejecting, throwing, rolling, lifting, or even catapulting the object from A to B. Thus, it is important to widen the solution space or system boundary as much as possible.

It is important to try to ensure that all the relevant inputs and outputs are listed clearly. As mentioned earlier, they can be classified as flows of materials, energy, and information.

Step 2. *Break down the overall function into a set of essential subfunctions. (Figure 5.5):* The conversion of the set of inputs into the set of outputs at the main function level is usually very complex. Thus, the black box needs to be broken into subtasks or subfunctions. The break-up depends on the experience of the designer, availability of components capable of performing specific tasks, and the latest technologies. In specifying subfunctions, it is necessary to express the task as a verb plus a noun (e.g., amplify signal, count items, decrease speed, increase diameter) or an action noun (e.g., actuator, loader, etc.). Each subfunction has its own input(s) and output(s).

Step 3. *Draw a block diagram showing the interactions between subfunctions:*

A block diagram consists of all subfunctions separately identified by enclosing them in boxes. They are linked by their inputs and outputs in order to satisfy the overall function of the product or device that is being designed. In other words, the original black box of the overall function is redrawn as a transparent box in which the necessary subfunctions and their links can be seen.

When drawing this diagram, we decide how the internal inputs and outputs of the subfunctions are linked so that we may make a feasible working system. We may find that inputs and outputs may have to be juggled and perhaps redefined, and some subfunctions may need to be reconnected. It is advised that different types of lines (continuous, dotted, and dashed lines) be drawn to indicate the different types of flows (material, energy, and information) within the block diagram.

Step 4. *Draw the system boundary:*

In drawing the block diagram, we also need to make decisions about the precise extent and location of the system boundary. For example, there can be no loose inputs or outputs in the diagram, except those that cross the system boundary.

The boundary now needs to be narrowed again because of its earlier broadening during the consideration of inputs, outputs, and overall function.

The boundary has to be drawn around a subset of the functions that have been identified in order to define a feasible product. It is also possible that the designer may not have complete freedom when drawing the system, since it may be restricted by the customer.

Step 5. *Search for appropriate components for performing each subfunction and its interactions:*

If the subfunctions have been defined adequately and at an appropriate level, then it should be possible to identify a suitable component for each subfunction. The identification of a component will depend on the nature of the product or device. For example, a component can be a mechanical, electrical, or combination device. A microprocessor-based electronic device can substitute some older versions of electromechanical devices. Since the function analysis method focuses on functions, new devices can be substituted at later stages of design or development.

5.5 FUNCTION STRUCTURE EXAMPLES

EXAMPLE 5.1 Grocery Cart

An example of function structure for a grocery cart is shown in Figure 5.6. There are three major functions, and each has several subfunctions. They can be summarized as follows.

1. Hold groceries
   a. Support weight
   b. Hold large items

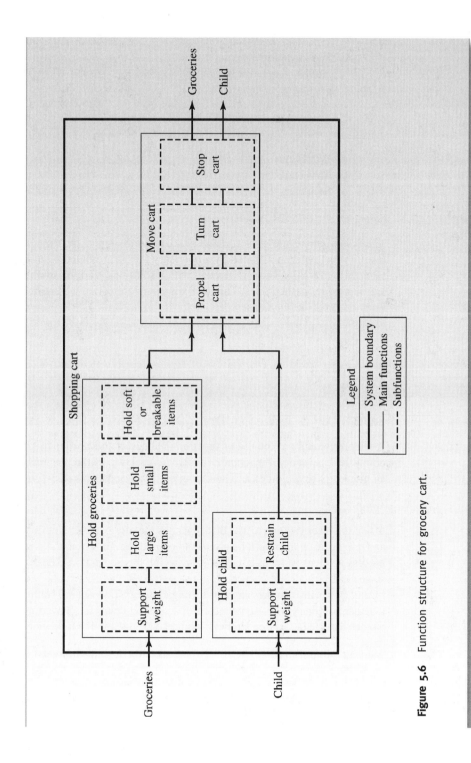

Figure 5.6 Function structure for grocery cart.

   c. Hold small items
   d. Hold soft or beverage items
2. Hold child
   a. Support weight
   b. restrain child
3. Move cart
   a. Propel cart
   b. Turn cart
   c. Stop cart

Solution

The system inputs are groceries and child, and the outputs are the same. The overall function is surrounded by a boundary that is presented by a solid line. Each of the functions also is blocked with a solid line, and the subfunctions are surrounded by dashed lines. This will keep the boundaries well defined for all functions and subfunctions.

EXAMPLE 5.2   Washer/Dryer (Adapted from Pahl & Beitz)

A simple example of the use of function analysis method is the design of the domestic washer/dryer machine. The overall function of the machine is to convert an input of soiled clothes into an output of clean clothes, as shown in Figure 5.7.

Solution

Inside the black box, a process that separates the dirt from the clothes must occur, and the dirt itself also must be a separate output. We know that the conventional process involves water as a means of achieving this separation. Thus, a further stage must be added to convert clean but wet clothes to clean and dry clothes. Even further stages involve pressing and sorting clothes. The essential subfunctions, together with the conventional means of achieving them (for converting soiled clothes into clean, pressed clothes), would be as given in the following chart.

Essential Subfunctions	Means of Achieving Subfunctions
Loosen dirt	Add water and detergent
Separate dirt from clothes	Agitate
Remove dirt	Rinse
Remove water	Spin
Dry clothes	Blow with hot air
Smooth clothes	Press

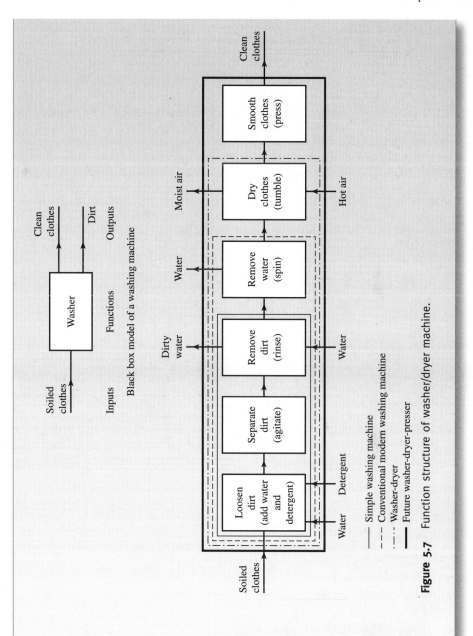

Figure 5.7 Function structure of washer/dryer machine.

The development of the washing machine involves progressively widening the system boundary, as shown in Figure 5.7. Early washing machines simply separated dirt from clothes but did nothing to remove excess water from the clothes. The process of wringing the clothes dry was left to the human operator. The inclusion of a spin-drying function removed excess water but still left a drying process. This process is now incorporated into a washer-dryer system. Perhaps the smoothing and ironing of clothes should be incorporated in future designs.

EXAMPLE 5.3   Newspaper Vending Machine

A design team was asked to develop a newspaper vending unit that is able to dispense three types of papers and not be easily vandalized. The design team proposed the following functional structure for the unit. The unit input is newspapers, and the output is a sold newspaper. The following are the functions and subfunctions as suggested by the design team.

1. Load the papers
2. Align the papers
3. Hold the papers
4. Payment
   a. Accept payment
   b. Count payment
   c. Payment trigger
5. Release a paper
6. Dispense a paper

Solution

Figure 5.8 shows the function structure for the system.

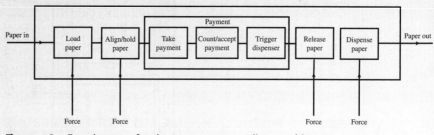

Figure 5.8   Function tree for the newspaper vending machine.

5.6   REVERSE ENGINEERING

Previous sections have discussed ways in establishing new products derived from a set of customer requirements. However, in many instances, the product already exists in the market (either your own product or a competitor's). In this case, an existing design is analyzed into subsystems which are subsequently analyzed in generic terms to ultimately establish the product concept or specification in solution-neutral form as the function tree. This tree and other data generated will then allow the designer to identify the weak subsystems and replace them or to use latent technological development when they prove to be advantageous. This process is commonly known as *reverse engineering* and is essentially functional decomposition in 'reverse'. It is important when a product is already available and the designers wish to benchmark, adapt, or improve the design.

Reverse engineering starts with the description of the process or the performing of the functions by the product. This exercise is carried out with the intention of identifying the underlying principles of the product and its subsystems. From the description of the process, it is possible to identify the subsystems and the constituent parts (a parts tree) of the product under consideration. This gives the *embodiment design* of the product. From this, it is possible to identify the functions performed by the different subsystems. The process in carrying out reverse engineering can be summarized by the following.

1. Description of the process.
2. Breaking down the product into subsystems.
3. Establishing the functions of the subsystems.

5.6.1 Reverse Engineering Example—Dishwasher
Process Description

Dirt is usually a mixture of fatty and solid particulate material (although solids may be present without fat) and will usually contain any or all of the following: pigment, carbon, clay, iron oxide, protein, and fat. When particulate dirt is present on its own, it can be removed by simply washing with water. When this particulate dirt is mixed fat however, water is not sufficient to remove the dirt. The fat adheres to the surface, and a chemical is required to reduce this clinging force and remove the dirt. The chemical must also prevent the dirt from redepositing on the surface. These chemicals are the detergents or soaps that are used for washing. Over the period of time, detergent manufacturers have tried to develop detergents that require lesser and lesser mechanical agitation.

There are two main processes used by detergents to remove dirt. The first method increases the contact angle of the fat with respect to the surface to which it is attached. This causes the fat to roll up into a ball, which then can be removed easily. The second process uses the polar chemical characteristics of sebum. Intermolecular attraction between a part of the detergent known as the surfactant and the fat results in the fat being more attracted to the detergent solution than to the surface. Hence, the fat leaves the surface and is held in the solution. The first of these two processes takes place at higher temperatures than the second. For this reason, it is clear that the washing process will need to take place at very specific temperatures. When a program is set, the detergent is flushed by the flow of water. Then the dishes and the water with the detergents are agitated and heated according to the pattern predetermined by the program selected. This ensures the removal of all dirt. The dirty water with the detergent is then pumped out, and fresh water is filled into the machine for rinsing purposes. Drying is then optionally applied by blowing hot air onto the dishes.

Functional Subsystems

From the description of the process, the following functional subsystems can be identified.

- Water addition system
- Detergent addition system
- Heating system
- Container and agitation system
- Dirty water removal system
- Rinsing system
- Drying system

Figure 5.9 illustrates the reverse engineering of the dishwasher, starting from the parts (embodiments) at the bottom of the chart and working up towards the functions. Refer back to Section 5.6.2 (the washing machine) and try to compare the function structures of

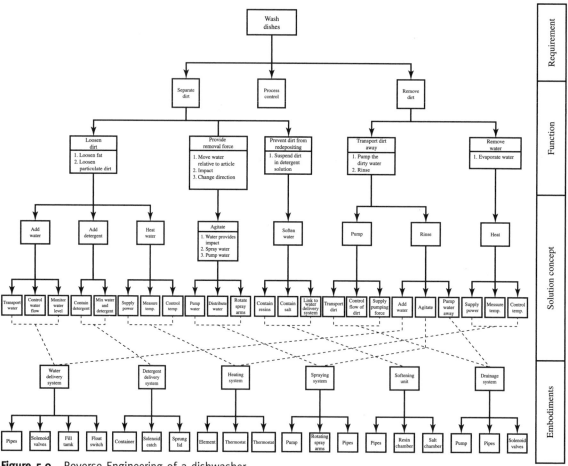

Figure 5.9   Reverse Engineering of a dishwasher.

each product. You will notice how alike they are. This demonstrates the power of remaining solution-neutral towards the beginning of a design, as it leaves your options to venture into adapted designs and different products (washing machine and dishwasher) still based around similar functions and concepts. In this case, many of the parts that perform similar functions in the two products can be standardized, which will lead you to a single core design and then to variants which differentiate between the two products. This is the basis for *design reuse* and is commonly used by companies wishing to 'diversify' their product offerings while remaining within their own expertise and cutting down costs by reusing many of their existing parts.

5.7 REVERSE ENGINEERING EXAMPLE—PAPER STAPLER

Now let us look at an example of determining the functional structure of another existing design. Consider a paper stapler. Students are encouraged to take apart an existing paper stapler and identify the different components. They should then identify the specific task of each of the components. Either individually or collectively, the components serve a specific function or subfunction, as seen in the Figure 5.10. Match your components to the listed functions.

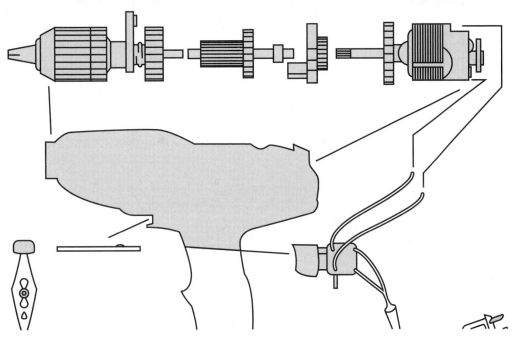

Figure 5.10 The engineer must design products, keeping in mind how those products will be manufactured. This lab introduces the concepts of designing for manufacturing, function analysis, and designing for assembly. Such reverse engineering techniques will not only help the engineer to produce a better product, but will help save manufacturing costs.

LAB 8: Reverse Engineering

Purpose

This lab introduces concepts such as design for manufacturing, function analysis, and design for assembly. In this lab you will disassemble a drill to its essential components and then assemble it back to its original shape.[1] Before you assume responsibility for the drill, make sure that it is in good working condition. The following types of drills are available (we used three types; one type is enough for this activity): Black & Decker DR210K, Skil Twist 2106, Skil Twist XTRA 2207.

Procedure

1. Identify the type of drill, manufacturer, model number, and performance specifications.
2. Plug in the drill and run it. Listen to how it sounds and feels as it runs. Your senses are a useful qualitative measurement tool. The drill should sound and feel the same after you take it apart and put it back together again.
3. Before you disassemble the drill, list all of the functional requirements that the drill must satisfy. For example, one function could be "hold a drill bit without slipping."
4. Unplug the drill and disassemble it as far as possible. *Hint: Carefully sketch as you disassemble*. Put all parts in a bin.
5. After listing all functional requirements and disassembling the drill, list the physical parts or assemblies that satisfy each function. In the preceding example the function "hold a drill bit without slipping" would be satisfied by a drill chuck. After disassembling the drill to its basic components, you may find out that you didn't list all functions. Make a table showing the functions and the corresponding part or assembly that satisfies that function.
6. Determine the gear ratio. What types of gears are used? What are some other gearing alternatives that could be used in this assembly? Why are they not used?
7. Determine the type of bearing used. Why do you think these types were chosen?
8. Determine what type of motor is used. Why do you think this type is chosen? Describe how the motor in your drill works in terms of basic physical principles. Use illustrations to help your description.
9. Make a bill of material (part list) of your drill. Identify the material used in each part. Also indicate whether the part is an off-the-shelf item or would need to be specifically manufactured for this product (i.e., custom part). What percentage of the total number of parts can be purchased from a parts vendor? Explain your findings.
10. What is the total number of parts in your drill? Compare this number to other drills being dissected. Are there any parts that can be eliminated or combined into a simpler part?
11. Observe the tolerances in the assemblies. Why do you think these tolerances are needed?
12. Reassemble your drill. Be especially careful of the motor brushes.
13. What features of your drill make it easy to assemble? Is there anything that was particularly difficult to reassemble? How would you improve the design of your drill to eliminate this

---

[1]This lab is based on material presented at an NSF workshop, Central Michigan University, 1999.

shortcoming? Are there any other improvements to the design that could improve the assembling of the drill? How long would it take for a factory worker to assemble the dull?

14. Once the drill is reassembled and you have no leftover parts, plug it in and try to run it. If it does not sound or feel like it did before you disassembled the drill, or if a strange smell develops, STOP IMMEDIATELY. Unplug the drill and try to determine what went wrong with your assembly steps.

Safety note: **Never plug the cord into an electrical outlet while the drill housing is open.**

Lab 8 Problems

Videotape the lab. Doing so provides visual reinforcement of the parts presented in the lab. ■

5.8   PROBLEMS

5.8.1   Team Activities

1. Discuss the following statement as it relates to the stage of the design process you have achieved thus far. The design process is the divergent–convergent process.

2. As was discussed in this chapter, there are several trends and philosophies on design. Perform a technical search and list four different trends.
   Discuss the differences and similarities among these trends.

3. Why would it be wrong to state a mechanism of performing a function rather than generalize and name the function in building up a function structure?

4. What is the difference between functions and subfunctions?

5. Would it be possible for different design teams to come up with different function structures for the same system/product? Why or why not?

6. As a team, choose one of the following products to take apart. Name the parts and what function they serve. Then build a functional structure for the product.
   a. Paper hole puncher
   b. Cellular phone
   c. Computer mouse
   d. Umbrella
   e. Tape dispenser
   f. Can opener

7. Develop a function tree to produce a device that is capable of converting wood chips into a fire log.
   a. Express the overall function for the system.
   b. Break down the over all function into a set of subfunctions.
   c. Draw a block diagram showing the interactions between the subfunctions.
   d. Draw the system boundary.

8. List the steps needed to generate the functional structure.

9. What is the difference between the functional structure and an objective tree?

10. Why can't you find a function for each objective you generate in the objective tree?

11. What function would you propose to satisfy the "must be safe" objective for the newspaper vending machine?

12. Generate a function tree for a coin sorter device.

13. Generate a function tree for a leaf-removing device.

5.9   Selected Bibliography

BIRMINGHAM, R., CLELAND, G., DRIVER, R., and MAFFIN, D. *Understanding Engineering Design.* Upper Saddle River, NJ: Prentice Hall, 1997.

BURGHARDT, M.D. *Introduction to Engineering Design and Problem Solving.* New York: McGraw-Hill, 1999.

CROSS, N. *Engineering Design Methods.* New York: Wiley, 1994.

CROSS, N., CHRISTIAN, H., and DORST, K. *Analysing Design Activity.* New York: Wiley, 1996.

DYM, C. L. *Engineering Design: A Synthesis of Views.* Cambridge, UK: Cambridge University Press, 1994.

EDER, W. E. "Problem Solving Is Necessary, But Not Sufficient." *American Society for Engineering Education (ASEE) Annual Conference*, Session 2330, Milwaukee, WI, 1997.

EDER, W. E. "Teaching about Methods: Coordinating Theory-Based Explanation with Practice." *Proceedings of the American Society for Engineering Education (ASEE) Conference*, Session 3230, Washington, D.C., 1996.

EEKELS, J. and ROOZNBURG, N. F.M. "A Methodological Comparison of Structures of Scientific Research and Engineering Design: Their Similarities and Differences." *Design Studies*, Vol. 12, No. 4, pp. 197–203, 1991.

FULCHER, A. J. and HILLS, P. "Towards a Strategic Framework for Design Research." *Journal of Engineering Design*, Vol. 7, No. 2, pp. 183–194, 1996.

JANSSON, D. G., CONDOOR, S. S., and BROCK, H. R. "Cognition in Design: Viewing the Hidden Side of the Design Process." *Environment and Planning B, Planning and Design*, Vol. 19, pp. 257–271, 1993.

HOLT, K. "Brainstorming from Classics to Electronics." *Proceedings of Workshop, EDC Engineering Design and Creativity*, edited by W. E. Eder, Zurich, pp. 113–118, 1996.

HUBKA, V., ANDREASEN, M. M., and EDER, W. E. *Practical Studies in Systematic Design.* London: Butterworths, 1988.

HUBKA, V. and EDER, W. E. *Theory of Technical Systems.* New York: Springer-Verlag, 1988.

HUBKA, V. and EDER, W. E. *Design Science: Introduction to the Needs, Scope and Organization of Engineering Design Knowledge.* New York: Springer-Verlag, 1996.

KUHN, T. S. *The Essential Tension: Selected Studies in Scientific Tradition and Change.* Chicago: University of Chicago Press, 1977.

PAHL, G. and BEITZ, W. *Engineering Design: A Systematic Approach.* New York: Springer-Verlag, 1996.

SCHON, D. A. *The Reflective Practitioner: How Professionals Think in Action.* New York: Basic Books, 1983.

SUH, N. P. *Principles of Design.* Oxford University Press, 1989.

Tuomaala, J. "Creative Engineering Design." *Proceedings of Workshop, EDC Engineering Design and Creativity*, edited by W. E. Eder, Zurich, pp. 23–33, 1996.

Ulman, D. G. *The Mechanical Design Process.* New York: McGraw-Hill, 1992.

Vidosic, J. P. *Elements of Engineering Design.* New York: The Ronald Press Co., 1969.

Walton, J. *Engineering Design: From Art to Practice.* New York: West Publishing Company, 1991.

CHAPTER •6

Specifications

At the specification stage of the design process, the engineer can continue to provide clarification of the needs statement. This rider needs to stay dry when biking through a wet forest. How can the bike's splashguard be designed to fill such a need? (Inc/Shutterstock)

6.1 OBJECTIVES

By the end of this chapter, you should be able to

1.   Quantify qualitative objectives.
2.   Utilize techniques to organize specifications into categories.
3.   Further clarify the need statement.

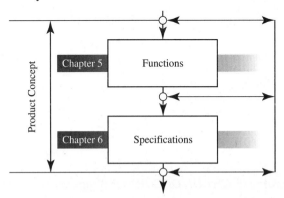

S pecifications are important to achieving a successful design. At the specification stage of the design process, you continue to provide additional clarification of the need statement. This is the last step in defining the problem before you begin to suggest possible solutions. So far, the objective tree provided a tool with which to list customer needs and design team objectives into categories. It defines the vague statements that are introduced by the customer. The function structure provided a mechanism with which to activate the objectives. However, the objective tree and the structural functions do not set specific limits on the different functions and objectives. The objective tree or function structures are statements of what a design must achieve or do, but they are not normally set in terms of precise limits, which is what a performance specification does. For example, an objective may be stated to develop a product which has low weight. Yet the term *low weight* is not clearly defined. Does it mean 20 kg or 100 kg?

A specification consists of a metric and a value. For example, "average time to assemble" is a metric, while "less than 75 seconds" is the value of this metric. Note that the value may take on several forms, including a particular number, a range, or an inequality. Values are always labeled with the appropriate units. Together, the metric and value form a specification. The product specifications are simply the set of the individual specifications. Metrics are usually derived from the function tree established in Chapter 5. This step allows you to define your vague and ill-measured objectives. One technique is to list a few ways you could measure an objective. Thus, you would brainstorm for metric-value combinations that are suitable in measuring an objective. For example, one of the objectives could be that the product is lightweight. Engineering specifications to measure this objective can be listed as

a.   Volume of the product should be less than [*value suitable to the product*].
b.   Mass of the product should be less than [*value suitable to the product*].
c.   Material density should be less than [*value suitable to the product*].

Some customers (10 to 20% on average) are able to provide specifications that set limits and clarify vague requirements. An example of such is an announcement from the National Inventors Council for the design of a stairclimbing wheelchair.[1]

NATIONAL INVENTORS COUNCIL
U.S. DEPARTMENT OF COMMERCE
WASHINGTON, D.C. 20550

There is a great need for a STAIR-CLIMBING WHEELCHAIR. The objective of such a chair is to give an active handicapped individual an additional range of mobility. The desired design should enable the individual to cope with the usual problems he or she might encounter when traveling to and from work, moving about industrial buildings, and the like. It must be remembered that the chair will perform the usual wheelchair function approximately 95% of the time; therefore, not too much of the conventional wheelchair's versatility and convenience should be sacrificed in providing the climbing function. In this connection, it might be well to point out that many active, handicapped people are able to fold a conventional wheelchair, put it in an automobile, and drive to work. If this cannot be accomplished with a climbing chair, the overall objectives of providing the handicapped with independent mobility will not be achieved.

The following factors are to be taken into consideration:

(a) *WEIGHT OF OCCUPANT: Assumed to be 100 kg*

(b) *WEIGHT OF CHAIR: 25 to 35 kg, maximum.*

(c) *COLLAPSIBILITY: Capable of being folded by the user and stowed in the interior of a standard automobile or taxi cab. Special loading ramps (which are not carried) as a part of the chair system or special modifications to the automobile to accommodate chairs will not be considered.*

(d) *WIDTH OF CHAIR: 25 inches/63 cm maximum. The American Standards Association utilized this width as the maximum for standard wheelchairs when establishing architectural standards, which provide the physically handicapped with freedom of movement in buildings and other structures. However, a lesser width for a chair or a means of temporarily reducing the full open width for passage through narrow doorways would provide a definite asset for negotiating door openings in private homes or other structures, which do not conform to the new standards.*

(e) *TURNING ABILITY: The turning radius should be held to a bare minimum. Turning should be accomplished without damage to floors or carpets. Chair must be able to negotiate an L or a U-type stair landing. Ability to negotiate an L-type landing as small as 1 m by 1 m would be considered an asset.*

(f) *CLIMBING ABILITY: The chair should be able to negotiate street curbing and any stair with average height risers and depth tread that is found in office buildings and homes. Ascent should be performed in the presence of litter and moisture without damaging stair treads and risers. Any transition*

---

[1]National Inventors Council, U.S. Department of Commerce

or adjustments required between normal and ascending and descending functions must be accomplished in a minimum time to eliminate delays at street curbs and stairways.

(g)   *PROPULSION SYSTEM: The chair may be propelled by the occupant or by a motorized unit; however, total cost, weight, and collapsibility requirements are the same for either type of drive. If self-driven, a minimum arm strength of 5 kg may be used in devising the drive mechanism.*

(h)   *CHAIR SAFETY: The presence of an attendant, although undesirable, is permissible during ascent or descent. However, the chair occupant should always be in a reasonable state of balance, so that no more than 8 to 14 kg of weight will be transferred to the attendant. The chair must be "fail safe" to prevent uncontrolled descent in the event something happens to the occupant, attendant, or mechanism.*

(i)   *COST: In order for the maximum number of handicapped persons to be able to purchase a stair-climbing wheelchair, every effort should be made to keep its cost at a bare minimum. The retail cost should be no more than $200. Current models of standard, tubular-frame folding wheelchairs are priced at approximately $150.*

(j)   *OTHER CONSIDERATIONS: Chair should be capable of performing all of its normal functions without undue jostling or jouncing of the occupant. Since most paraplegics have little control of the lower limbs, a method of retaining them in position is often required. The footrests should be movable in order to enable the occupants to raise themselves to a standing position on the floor.*

(k)   *The chair should not require the installation of special ramps, mechanical contrivances, or electrical outlets in buildings.*

The preceding example has a clear statement with a clear set of requirements. Therefore, an objective tree, a function structure, and even the market analysis may not require much effort, because all necessary details and specific requirements are spelled out in the statement. This is not always the case; 80 to 90% of the time the statement given to the design engineer is very vague. For instance, it may simply say, "Design a stair-climbing wheelchair." In this case, an objective tree, functional analysis, and market analysis are essential to clarify the statement. The ability to write a good set of engineering specifications is proof that the design team understands the problem.

There are many techniques for generating engineering specifications. Hubka separates the properties affected by the constraints into categories based on operational, ergonomic, aesthetic, distribution, delivery, planning, design, production, and economic factors. Pahl and Beitz based the specifications on the fulfillments of the technical functions, the attainment of economic feasibility, and the observance of safety requirements for the human and environment.

The approach that will be followed in this book is to develop a set of requirements based on the performance-specification method and the quality-function-deployment method. In the performance-specification method, the specification defines the required performance instead of the required product. It also describes the performance that a design solution has to achieve but not any particular physical component that may constitute a means of achieving that performance. In the quality-function-deployment method, the attributes of a required performance are translated into engineering characteristics.

6.2 PERFORMANCE-SPECIFICATION METHOD

The following are the steps in producing a set of specifications using the performance specifications method.

Step 1. *Consider the different levels of generality of solution that might be applicable:*
Specifications that are set at too high a level of generality may allow inappropriate solutions to be suggested. However, specifications which are too tight may remove all of the designer's freedom and creativity for the range of acceptable solutions. This level of limitation may also be connected with the definition of the customer you are dealing with, because the customer may also limit the class and the set of specifications. For example, designing a space shuttle puts the designer on a tighter limit of specification. Another example that demonstrates this point involves the design of an air jet for civilian use. Such a design has a different set of requirements than designing an air jet for military use. Recognizing the customer's needs is an important factor in defining the range for the set of requirements. The level of generality could be reserved through the development of the objective tree.

Different types of generality levels can be listed from the most general to least general and may be demonstrated as

a. Product alternatives
b. Product types
c. Product features

Let us consider an example to illustrate these levels. Suppose that the product in question is a domestic aluminum-can disposal device. At the highest level of generality, the designer would be free to propose alternative ways of disposing of aluminum cans, such as crushing, melting, shredding and chemical. There might even be freedom to move from the concept of one can to multican or from domestic to factory. At the intermediate level, the designer would have much more limited freedom and might only be concerned with different types of can crushers, such as foot operated or automatic. At the lowest level of generality, the designer would be constrained to consider only different features within a particular type of crusher, such as if it hangs on the wall or is self-standing, processes one or multicans in one step, and so on. Thus, at this level of developing specifications, designers need to review the need statement and objective tree in order to analyze where the objective tree stands with respect to the mentioned levels.

Step 2. *Determine the level of generality at which to operate:*
In general, the customer/client determines the level of generality at which the designer may operate. For example, the customer may ask the designer to design a safe, reliable aluminum-can crusher. Here, the customer has already established the type of product, and constraints are placed on the type of aluminum can disposable unit: It must be a crusher. Hence, the set of requirements will follow for a can crusher, not a disposable can unit. It should be noted however that sometimes customers may state a solution to a product such as 'can crusher', where in fact, they do not really care whether the cans are crushed or not, but they actually just need a product that disposes cans efficiently. Therefore, it is important to communicate with the customer to ensure that the apparent limits imposed by them are really necessary, as they may have simply not considered other alternatives. This further emphasizes the need to accurately define the problem before suggesting possible solutions.

Step 3. *Identify the required performance attributes:*

For clarifying the customer statement, objective trees, function analysis, and market analysis are used. Objective trees and functional analysis determine the performance attributes of the product. Some of the requirements in the customer statement are "must meet" (called demands) and others are "desirable" (called wishes). Solutions that do not meet the demands are not acceptable solutions, whereas wishes are requirements that should be taken into consideration whenever possible. Pahl and Beitz developed a checklist that can be used to derive attributes along with objective trees and function analysis (Table 6.1). Once the attributes are listed, distinguish whether the attribute is a demand or a wish, and then tabulate the requirements.

TABLE 6.1 Checklist for drawing up a requirement list.[2]

Main Headings	Examples
Geometry	Size, height, breadth, length, diameter, space, requirement, number, arrangement, connection, extension
Kinematics	Type of motion, direction of motion, velocity, acceleration
Forces	Direction of force, magnitude of force, frequency, weight, load, deformation, stiffness, elasticity, stability, resonance
Energy	Output, efficiency, loss, friction, ventilation, state, pressure, temperature, heating, cooling, supply, storage, capacity, conversion
Materials	Physical and chemical properties of the initial and final product, auxiliary materials, prescribed materials (food regulations, etc.)
Signals	Inputs and outputs, form, display, control equipment
Safety	Direct safety principles, protective systems, operational, operator and environmental safety
Ergonomics	The man–machine relationship, type of operation, clearness of layout, lighting, aesthetics
Production	Factory limitations, maximum possible dimensions, preferred production methods, means of production, achievable quality and tolerances
Quality	Control possibilities of testing and measuring, application of special regulations and standards
Assembly	Special regulations, installation, siting, foundation, transport limitations due to lifting gear, clearance, means of transport (height and weight), nature and conditions of dispatch
Operation	Quietness, wear, special uses, marketing area, destination (for example, sulphurous atmosphere, tropical conditions)
Maintenance	Servicing intervals (if any), inspection, exchange and repair, painting, cleaning
Recycling	Reuse, reprocessing, waste disposal, storage
Costs	Maximum permissible manufacturing costs, cost of tooling, investment and depreciation
Schedule	End date of development, project planning and control, delivery date

[2]From ENGINEERING DESIGN: A SYSTEMATIC APPROACH by G. Pahl and W. Beitz, transalated by Ken Wallace, Lucienne Blessing and Frank Bauert, Edited by Ken Wallace. Copyright © Springer-Verlag London Limited 1996. Reprinted by permission.

Step 4. *State succinct and precise performance requirements for each attribute:*
The attributes that can be written in quantified terms must be identified. The performance limit should be attached to the attribute. The limits may be set by the customer statement or by federal and government agency standards. These attributes may include, but are not limited to, maximum weight, power output, size, and volume flow rate. For presenting the specifications, use a metric-value combination (i.e., size less than $1\,m^3$). For wishes and demands that do not have defined value attached to the metric, we use the quality-function-deployment method, which is described in Section 6.4.

6.3 CASE STUDY SPECIFICATION TABLE: AUTOMATIC CAN CRUSHER

Continuing our can crusher example from previous chapters, the next stage is to draw up a specifications table. The prioritized requirements and market analysis results should be reviewed at this time (Chapters 3 and 4). From the need statement for the aluminum can crusher, the specifications that *must* be met are

1. The design of the crushing mechanism is not to exceed $20 \times 20 \times 10\,cm$ in the total size.
2. The can crusher must have a continuous feed mechanism.
3. The can must be crushed to 1/5 of its original volume.
4. The device must operate safely. Children will use it.
5. The device is fully automatic.

Table 6.2 shows an example specification table that includes a metric and a value for this case study.

TABLE 6.2 Specification table for automatic can crusher.

Metric	Value
Dimensions	$20 \times 20 \times 10$ cm
Cans crushed	1/5 original volume
Weight	< 10 kg
Sales price	$< \$50$
Number of parts	< 100
People able to use	> 5 yrs
Probability of injury	$< 0.1\%$
Manufacturing cost	$< \$200$
Steps to operate	1
Maintenance cost	$< \$10$ annually
Efficiency rating	> 95 percentile
Internal parts enclosed	100%
Storage of crushed cans	60
Loader capacity	> 30 cans
Crush cans	≥ 15 cans/min

Crush cans	$\geq 1.2 \times 10^{-2}$ m^3/min
Crush cans	≥ 0.57 kg/min
Noise output	> 30 dB
Starts	< 10 sec
Runs	> 2 hours at a time
Stops	< 5 sec
Vibration magnitude	< 5 mm
Vibrations	< 4 sec
Withstand	≤ 250 N
Maintenance	< 4 hrs/yr
Number of colors	6
Lifespan	> 4000 hrs

6.4   QUALITY-FUNCTION-DEPLOYMENT METHOD

The quality-function-deployment (QFD) method was first developed in Japan in the mid-1970s and used in the United States in the late 1980s. Through the use of the QFD method, Toyota was able to reduce the cost of bringing a new car model to the market by more than 60%, and the time was reduced by about 33%.

This method brings together the work carried out in the previous chapters and allows each stage of the design process to be measured quantitatively on how well it is achieving the previous stage and, hence, how good the design is. A house of quality chart is drawn to measure specifications against initial requirements. Later another house of quality chart will be drawn to measure the conceptual design against the specifications and so on. This 'cascading' method (Figure 6.1) allows a follow through of the entire design process to measure how each stage of the design process addresses the initial requirements set by the customers.

Figure 6.2 shows the QFD house of quality chart 1, which rates specifications against the initial product requirements. The method to complete this chart is as follows.

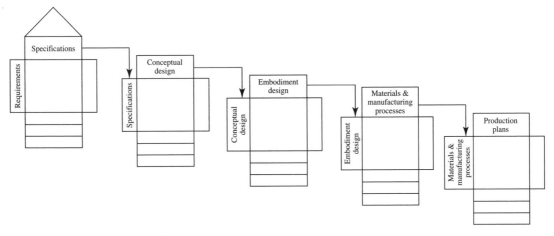

Figure 6.1   QFD charts cascade.

Region 1 The prioritized requirements established in Chapter 4 are listed as rows along with their importance ratings (1 to 9, 9 being the most important). You have done this through question/answer sessions with the customer and through discussion sessions with the design team to develop the objective tree. Market research, function analysis, and the performance-specification method also help in this effort. In cases where a company is launching/designing a new product where there is no particular customer at hand, gathering attributes could be done through direct interaction with prospective customers by conducting

- Product clinics: Customers are quizzed in depth about what they like and dislike about a particular product. User surveys also can be used here.
- Hall test (tests conducted in the same hall): Various competing products are arranged on display, and customers are asked to inspect the products and give their opinions.

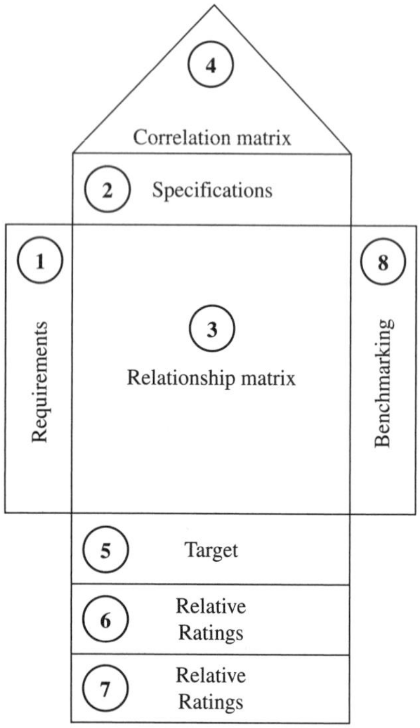

Region 2 Specifications established in this chapter are listed as columns.

Figure 6.2   Stage 1 QFD house of quality chart.

Region 3 Each specification is then rated as a CORRELATION to each requirement. This is to find out how well each specification addresses each requirement. If there is NO correlation, the grid space is left blank. If there is a slight or weak correlation, rate as 1. If there is medium correlation, rate as 3. If there is high/strong correlation, rate as 9. ONLY blank, 1, 3, or 9, are valid options in the relationship matrix region.

Region 4 Engineering specifications may have relationships between each other. For example, a powerful engine is also likely to be a heavier engine. This interaction is added as a roof to the matrix. Region 4 is this correlation matrix. It also identifies specifications that are in conflict with each other. Once again, correlation ratings of 1, 3, or 9 are used, but in addition to this and the blank, if there is a conflict, then a ' − ' sign should be placed between the conflicting specifications.

Region 5 Depicts target values for the specifications to improve over competitors: The market analysis may be important at this stage to identify the market limits.

Region 6 The absolute importance ratings of the specifications as measured against the prioritized requirements. This is achieved by multiplying each specification rating by its corresponding requirement importance rating and adding up the respective columns to get the absolute importance rating for that specification.

Region 7 The relative importance ratings and these values are the absolute importance ratings weighted relative to each other. Here, the highest absolute rating becomes the benchmark value and is given a relative importance of 9. All other specifications are then compared to this value.

Therefore, in Figure 6.3, there are four requirements (Safe, reliable, low cost, and pleasing appearance). Assume that these requirements derive five specifications, and these are mapped in the columns as shown in the chart. The calculation for the absolute importance rating of specification 1 is

$$(1 \times 9) + (1 \times 7) + (9 \times 2) + (3 \times 5) = 75$$

At first glance it may seem that specification 1 is the most important specification, as it is relevant to all of the requirements in some way. However, it is also somewhat unsafe (the most important requirement according to the importance rating). Specification 2 however is only relevant to one requirement (safety), but it highly correlates with this requirement. As a result, it yields an absolute importance rating of 81. This becomes the important specification to focus on, followed by specification 1. Specification 4 is the least important specification, as it only focuses on low cost and good looks but has nothing to do with the safety or reliability of the product. As you can see from this simple example, the QFD chart focuses you on the most important specifications.

Since specification 2 has the highest absolute importance rating, nine becomes the relative importance rating, and all other specifications are weighted down against this specification. Hence, the relative importance rating for specification 2 would be calculated by $(75/81) \times 9 = 8$ (rounded down to the nearest whole number).

Region 8 The benchmark value of each requirement is measured against competing products in the market. The objective here is to determine how the customer perceives the competition's ability to meet each of the requirements. Usually, customers make judgments about the product in terms of comparison with other products. This step is very important because it shows opportunities for product improvement. The market analysis you have conducted should play a vital role in this step. In many instances, students who do not conduct extensive market analysis find it hard to complete this step accurately.

	Importance rating	Specification 1	Specification 2	Specification 3	Specification 4	Specification 5
Safe	9	1	9	3		
Reliable	7	1			3	
Low cost	2	9		9	3	3
Pleasing appearance	5	3				3
Target information						
Absolute importance		75	81	45	27	21
Relative importance		8	9	5	3	2

Figure 6.3 Simple QFD Chart

Kano Model

In QFD, consumers are the most important factor, and the design should be arranged in a manner to meet their satisfaction. The Kano model can be used to measure customer satisfaction. In the Kano model, customer satisfaction is measured by the product function, as shown in Figure 6.4.

The Kano model was developed by Dr. Noriaki Kano in the early 1980s. In the Kano model, there are three different types of product quality that give customer satisfaction: basic quality, performance quality, and excitement quality. With basic quality, customers' requirements are not verbalized, because they specify assumed functions of the device. The only time a customer will mention them is if they are missing. If they are not fully implemented in the final product, the customer will be disgusted with it. If they are included, the customer will be neutral. An example is the requirement that a bicycle should have brakes. The performance quality refers to customers' requirements that are verbalized in the form that indicates the better the performance, the better the product. The excitement quality involves those requirements that are often unspoken, because the customer does not expect them to be met in the product. However, if they are absent, customers are neutral. If the customers' reaction to the final product contains surprise and delight at the additional functions, then the product's chance of success in the market is high.

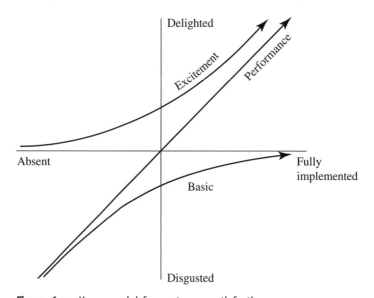

Figure 6.4   Kano model for customer satisfaction.

EXAMPLE 6.1   Controlled Vents

The design team in charge of designing an adjustment to current vents such that the vents could be opened and closed remotely from a centralized location in the house has developed a house of quality as shown in Figure 6.3. In this figure, the specifications are listed in the left column. The designers were the only people to evaluate the specifications. The designers developed engineering characteristics to measure the specifications as shown in Figure 6.5.

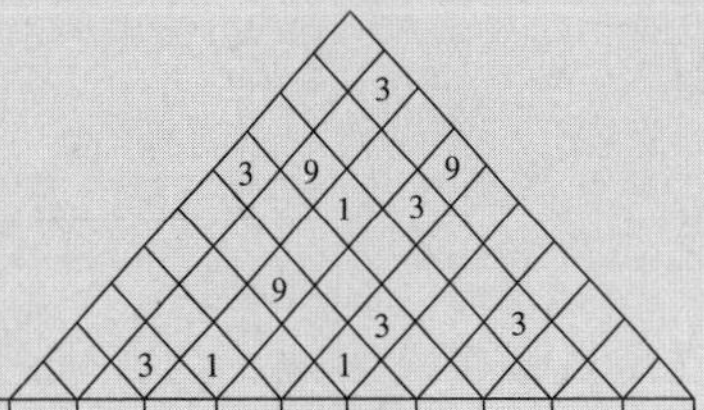

	Designer	Noise <30 db	Profit 25%	Parts <15	Steps to operate <5	Maintenance cost <$20 annually	Disassembly <15 minutes	Force <15 newtons	Production cost <$200 per unit	Time to open vent <5 seconds	Cost to use <$5 month
Easy to operate	9				9						
Low maintenance	3		1	9		9	3			1	
Few parts	5			9	3		3				
Safe for user	11	3			3						
Safe for environment	9	3									3
Enough power to do job	4				3			9		3	
Fits vents universally	4		9								
Low noise	2	9		3	3						
Low operation cost	4		3			1			3		9
Inconspicuous	1	3		3							
Profitable	8		9	3		3			9		3
Easy to disassemble	2			3			9				
Low vibration	2	3			3			9		3	
Long lasting	2			9	3			3		3	
Small actuating force	1							3		3	
Time in use small	2	3		3	9					9	
Inexpensive material	4		9			3			9		
Inexpensive to consumer	4		1	3		3			9		3
Conserve energy	2			1	3			3			9
Easy to repair	2			9		3	9				
Quality materials	3		3								
Different settings	4		3		3		3				3
Remotely operated	12		9	9			1		3		
Total	100										

Figure 6.5 House of quality for controllable vents. (Example 6.1)

EXAMPLE 6.2 Splashguard[3]

This example is of a relatively simple product. It illustrates that considerable effort may be necessary in designing to satisfy customer requirements even for a simple product. The design team identified three important customers for the product: the rider, the mechanic, and the marketing. The mechanics are considered customers, because they will sell (and perhaps install) the splashguard. Marketing is added to the customer list to ensure capture of production and

[3]MECHANICAL DESIGN PROCESS by Ullmann. Copyright 1996 by MCGRAW-HILL COMPANIES, INC.—BOOKS. Reproduced with permission of MCGRAW-HILL COMPANIES, INC.—BOOKS in the format Textbook via Copyright Clearance Center.

Preliminary list of customers'
requirements for the splashguard

Riders' and bike shop mechanics' requirements
    Keeps water off rider
    Is easy to attach
    Is easy to detach
    Is quick to attach
    Is quick to detach
    Won't mar bicycle
    Won't catch water/mud/debris
    Won't rattle
    Won't wobble
    Won't bend
    Has a long life
    Won't wear out
    Is lightweight
    Won't rub on wheel
    Is attractive
    Fits universally
    If permanent piece on bike, then is small
    If permanent piece on bike, then is easy to attach
    If permanent piece on bike, then is fast to attach
    If permanent piece on bike, then is noninterfering
    Won't interfere with lights, rack, panniers, or brakes

Company management requirements
    Captial expenditure is less than $15,000
    Can be developed in 3 months
    Can be marketable in 12 months
    Manufaturing cost is less than $3
    Estimated volume is $200,000 per year for 5 years

Figure 6.6 Requirements for splashguard. (Example 6.2) (Reprinted by permission of Pacific Cycle.)

marketing needs. Figure 6.6 shows a mountain bicycle and the customer's list of requirements for the splashguard.

Solution

Figure 6.7 shows the house of quality that is developed based on the QFD method for a splashguard. Try to calculate the absolute and relative importance ratings to identify the more important specifications. See if this matches your expectations of looking at them 'at first glance.'

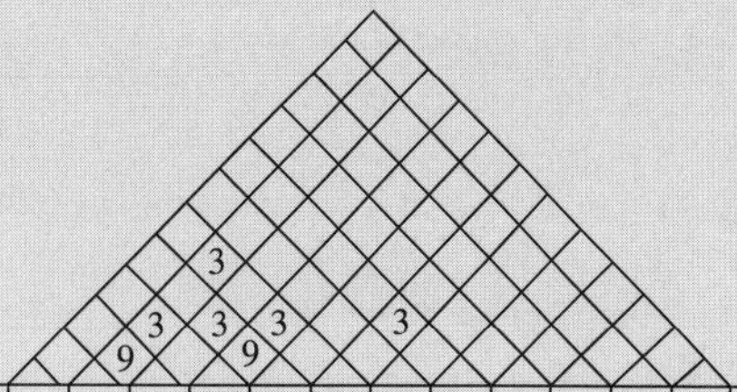

	Mechanic	Marketing	Rider	Water hitting rider (%)	Steps to attach (#)	Time to attach (sec)	Steps to detach (#)	Time to detach (sec)	Number of parts (#)	Weight (g)	Customers finding it visually appealing (%)	Color available (#)	Bikes it fits (%)	Upward release force (N)	Sales price ($)	Whale tail	Norco	Raincoat
Functional performance																		
Keeps water off rider	1	1	7	9										3		1	4	2
Fast to attach	5	4	8		3	9			3	1						1	4	3
Fast to detach	9	5	10				3	9	3	1						2	4	3
Can attach when dirty	7	13	12		3	3										3	3	2
Can detach when dirty	11	12	13				3	9								3	3	2
Human factors																		
Easy to attach	4	6	9	9					3	1						1	3	3
Easy to detach	10	7	11				9		3	1						1	4	3
Looks fast	2	10	2								9					4	2	2
Color matches bike	12	11	5								3	9				3	2	2
Interface with bike																		
Fits bike	3	3	3										9			3	2	4
Does not mar bike	8	8	6											1		3	1	4
Lightweight	6	9	4							9						3	3	4
Competitive sales price	13	2	1												9	2	3	1
Whale Tail				25	5	25	2	5	6	130	75	5	94	5	12			
Norco				0	3	5	1	3	2	140	65	1	65	15	15			
Raincoat				30	3	10	3	10	1	100	35	4	100	0	20			
Target				0	1	2	2	3	2	130	85	5	95	5	10			

Figure 6.7   House of quality for the splashguard. (MECHANICAL DESIGN PROCESS by Ullmann. Copyright 1996 by MCGRAW-HILL COMPANIES, INC.—BOOKS. Reproduced with permission of MCGRAW-HILL COMPANIES, INC.—BOOKS in the format Textbook via Copyright Clearance Center.)

6.5 HOUSE OF QUALITY: AUTOMATIC CAN CRUSHER

The design team has taken the wishes and objectives from the specification table and the objective tree and placed them into the house of quality. Because of the large number of wishes they listed on their table, they were unable to show the house of quality in one figure. However, four tables are shown that represent the house of quality (Figures 6.8 and 6.9). On the first table in Figure 6.8, the wishes are placed with the engineering characteristics, while on the second table, the engineering characteristics

	Probability of injury < 0.1%	Weight < 30 lb	Sales price < $50.00	Number of parts < 100	Dimensions (inch)	Crushing force > 30 lb	People able to use > 5 yrs	Manufacturing cost < $200	Steps to operate (1)	Maintenance cost < $10 annually	Efficiency rating > 95 percentile	Internal parts enclosed (100%)
Easy access to clear jammed cans	3			6			3	3		3	6	
Crushing mechanism humanly inaccessible	9	1	3	3	3		9	3				9
Machine rendered inoperable when opened	9		1	6			9	9				1
Easily accessible kill switch	9		3	7			9	3	3			
Ability to stop in mid operation	9		1	3			9	6	1			
No flying debris	9						9			1		6
Runs on 110V standard U.S house outlet			6	3		3	9	9	6		9	
Easy to start			3				9	6	6			
Long running capability			6	1			7			9	9	
High efficiency engine			6	1		9		9		9	9	
High material strength		9	6			3		9		6		
Small force required to depress the switches	9						9	3				
Safety stickers	9						9	2	2			
Machine reset button after kill switch activated	9			6			9	3	2			
Operation steps sticker				6			9	3	9			
Parts easy to acquire			6				3	9		9		
Internal parts safe from liquid damage	2		4	4				3		9	6	9
Crush many cans/min.			9	6	1	9	3	9			6	
No sharp corners	9						9	6				
Stops easily and immediately	9			3			9	6	1	6		
Low vibration		6	3	3	6			3		1		
Low peripheral force		6	3	3	6			3		1		
Shock absorption		6	3	3	6			3		1		
Utilize gravity in design		9	3	6	9			9			1	
No exhaust	9		1	1			9					
Low maintenance cost			6	6			8	3		9	6	1
Low noise output	9		4	3			6	6				9
Starts up immediately			3	5				3			6	
Low loading height	6		1			9	9					
Easily accessible interior	3		3				6	3		3		3
No service required			9	2			6			9	9	
Sealed bearings								3				6
Limit number of tight tolerances												
Low operation cost			9	3			6	3		6	6	
Easy access to clear jammed cans	6		3				3	3		6	3	3
Little heat produced	1			1		2					9	
Extra wiring insulation	6	1		1				1				
Enough force to crush glass, plastic						9	6	9				
Inconspicuous			6		6		6	3				

Figure 6.8A House of quality for automatic can crusher.

	Probability of injury < 0.1%	Weight < 30 lb	Sales price < $50.00	Number of parts < 100	Dimensions (inch)	Crushing force > 30 lb	People able to use > 5 yrs	Manufacturing cost < $200	Steps to operate (1)	Maintenance cost < $10 annually	Efficiency rating > 95 percentile	Internal parts enclosed (100%)
Consolidate mechanical functions	3			9				6	2		9	
Low standby power drain		1					3				9	
Aesthetically pleasing/blends with surrounding		9	2	6				8	6			
Utilizes ground to stabilize				1								
Less than five assembly steps			3	6				3	1			
Ability to mount to various surfaces	3	6					3	3	3			
Large can capacity loader	3	6					3	3	3	6		
Portable	9	6					3	6	3			
Durable refuse container								3				
Retail for < $50.00			9					9	9			3
Variable length/retractable cord			3					2	1			
Large storage of crushed cans	1		4				3	2	1	6		
Ability to crush various sizes of containers			9	3	8		6	9				
Easy to disassemble			3	6			3	3		6		
Easy cleaning			6				3	1		3		6
Green light to indicate ok to load	9		6					9	2			
Red light to indicate the crushing mechanism is in operation	9		6					9	2			
Yellow light to indicate improper use of the machine	9		6					9	2			
Automatically switches to standby power when not in use										6		
Receiving container on caster		2										
Weather proof			1	3				3		6		9
Crushes glass, plastic and aluminum containers			9	6	3	9	7	9				
Drain for residual liquid from machine	1			1				1		6		
Built from a polymer		6	6					6				
Housing constructed from a formed polymer		6	6					6				
Can counter				3	2			1	2			
Container to hold refuse liquid		2										
Flip open lid												
Colors available			9					3				
Paintable surface			8					2				
Plexiglas window to view operation			8					2				6

Figure 6.8B House of quality for automatic can crusher.

are evaluated against each other. The engineering characteristics are obtained by including a metric, a value, and a unit. Numerical values between 1 and 9 are assigned between each wish and engineering characteristic to rank the correlation between them; 1 is a very low correlation and 9 is a very high correlation. Blank spaces correspond to no correlation.

Storage of crushed cans (60 cans)	Loader capacity > 30 cans	Crush >= 15 cans/min.	Crush >= 1.2×10^{-2} m³ cans/min.	Crush >= 0.57 kg/min.	Noise output < 30 db	Starts < 10 sec	Runs < 2hrs at a time	Stops < 5 sec	Vibration magnitude < 5 mm	Vibrations < 4/sec	Withstand <= 250 N	Maintenance < 4hrs/yr	# colors (6)	Lifespan > 4000 hrs
		6	6	6								3		3
								1						
								9						
								9						
		1	1	1				9						
		6	6	6	3		1							
						9								
	1	2	2	2	3		9		2	1		9		9
		6	6	6	2				2	2		9		9
		1	1	1								6		9
						6								
												9		
												6		9
6	9	9	9	9	9	1	1		3	3				3
								9						
		9	9	9	9				9	9				3
		9	9	9	9				9	9				3
		9	9	9	9				9	9				3
					3				1	1				3
					1									
							6		6	6	7	9		9
		1	1	1	9		1		3	3				
		1	1	1		9								
	9													
		3	3	3				1				9		6
							3				6	9		9
												9		6
		3	3	3										
							6					9		3
		3	3	3				6				6		4
							3		1	1		1		4
														2
		9	9	9								9		
6	6								6	6			9	

Figure 6.8C House of quality for automatic can crusher.

Figure 6.9 Engineering characteristics.

	Probability of injury < 0.1%	Weight < 30 lb	Sales price < $50.00	Number of parts < 100	Noise output < 40 db	Dimensions (inch)	Cans crushed (20/min)	Crushing force > 30 lbs	People able to use > 5 yrs	Manufacturing cost < $200	Steps to operate (1)	Maintenance cost < $10 annually	Efficiency rating > 95 percentile	Internal parts enclosed (100%)	Storage of crushed cans (60 cans)	Loader capacity > 30 cans	Starts < 10 sec	Runs > 2 hrs at a time	Stops < 5 sec.	Vibration magnitude < 5 mm	Vibration < 4/sec	Withstand <= 250 N	Maintenance < 4 hrs/yr	# colors (6)	Lifespan > 4000 hrs
Probability of injury < 0.1%	9	6		9			3	9	3	3										3	3				
Weight < 30 lb		9		6	1					3						3	6			3	3				
Sales price < $50.00	6		9	3	6		9	3	9	9	6	6	9				6			3	3		9	9	9
Number of parts < 100		6	3	9	3					9		9	9			3							9		9
Noise output < 40 db	9		6	3	9		6		9																
Dimensions (inch)		1				9				3					6	6									
Cans crushed (20/min)		9		6			9	9					9				9	9	6				3		
Crushing force > 30 lbs	3	3					9	9					9							6	6				
People able to use > 5 yrs	9	9	9					9	9	6	6					9									
Manufacturing cost < $200	3	3	9	9		3				9				6	3	6								6	
Steps to operate (1)	3	6							9		9		9												3
Maintenance cost < $10 annually		6	9						6			9	9					6		3	3		9		
Efficiency rating > 95 percentile		9	9			9	9	6		6	9	9	9	6	6	6	6	6	6	6	6		9		9
Internal parts enclosed (100%)								6						9									9		
Storage of crushed cans (60 cans)		3				6		3					6		9	6		6							
Loader capacity > 30 cans		6				6	9			6			6		6	9	9								
Starts < 10 sec			6	3			9		9				6				9								
Runs > 2hrs at a time							6					6	6			6	9	9							6
Stops < 5 sec													6						9						6
Vibration magnitude < 5 mm	3	3	3					6				3	6							9					6
Vibration < 4/sec	3	3	3					6				3	6								9				6
Withstand <= 250 N																						9			
Maintenance < 4 hrs/yr			9	9		3						9	9	9									9		9
# colors (6)		9							6															9	
Lifespan > 4000 hrs			9	9						3			9				6	6	6	6			9		9

6.6 PROBLEMS

6.6.1   Team Activities

1. Given the need statement "Design a domestic heating appliance," generate different levels of generality of solutions.
   a. Alternatives
   b. Types
   c. Features

2. Generate engineering characteristics for the following:
   a. Easy to hold
   b. Cost effective
   c. Safe for the environment
   d. Easy to maintain

3. Consider the statement you worked on previously during an in-class team activity. By looking at your objective tree, generate a list of specifications and house of quality for the following statement.

   If you drive in the state of Florida, you notice some traffic congestion due to trees being trimmed on state roads and freeways. Assume that the

Florida Department of Transportation is the client. Usually, branches 6 inches or less in diameter are trimmed. The allowable horizontal distance from the edge of the road to the branches is 2m. The material removed from the trees must be collected and removed from the roadside. To reduce the cost of trimming, a maximum of only two workers can be assigned for each machine. The overall cost, which includes equipment, labor, etc., needs to be reduced by at least 25% from present cost. The state claims that the demand for your machines will follow the price reduction (i.e., if you are able to reduce the cost by 40%, the demand will increase by 40%). Allowable working hours depend on daylight and weather conditions.

6.6.2   Individual Activities

1. Develop a list of specifications for the following objectives. The table should include metrics, the corresponding values (assume a value), and units.
   a. Easy to install
   b. Reduces vibration to the hand
   c. Is not contaminated by water
   d. Lasts a long time
   e. Is safe in a crash
   f. Allows easy replacement of worn parts

2. List a set of metrics corresponding to the need that a pen writes smoothly.

3. Devise a metric and a corresponding set for the need that a roofing material lasts many years.

4. How might you establish precise and measurable specifications for intangible needs, such as "The front suspension looks great"?

5. Can poor performance relative to one specification always be compensated for with high performance on the other specifications? Give examples.

6. Develop a house of quality for the following need statement.

   We have a mountain-sized pile of wood chips that we want processed into home use fire logs. We are located in Tallahassee, Florida. We can have a continuous supply of wood chips throughout the year. We should be able to produce 50 fire logs per minute. The shareholders require us to have a large profit margin, and our prices should be lower than our competition for the same size logs. Needless to say, the logs should produce enough heat to keep our customers faithful and have environmentally clean exhaust.

7. Use a Web search engine to identify a list of companies that utilize QFD. Describe what they apply QFD for and list differences and similarities in the steps developing QFD.

6.7 Selected Bibliography

AKAO, Y. *Quality Function Deployment: Integrating Customer Requirements into Product Design.* [Translated by Glenn Mazur]. Productivity Press, 1990.

CLAUSING, D. and PUGH, S. "Enhanced Quality Function Deployment." *Design and Productivity International Conference,* Honolulu, HI, February 6–8, 1991.

COHEN, L. *Quality Function Deployment: How to Make QFD Work for You.* Reading, MA: Addison-Wesley, 1995.

DAETZ, D. *Customer Integration: The Quality Function Deployment (QFD) Leader's Guide for Decision Making.* New York: Wiley, 1995.

DAY, R. G. *Quality Function Deployment: Linking a Company with Its Customers.* ASQC Quality Press, 1993.

DEAN, E. B. "Quality Function Deployment for Large Systems." *Proceedings of the 1992 International Engineering Management Conference,* Eatontown, NJ, October 25–28, 1992.

HAUSER, J. R. and CLAUSING, D. "The House of Quality." *The Harvard Business Review,* May–June, No. 3, pp. 63–73, 1988.

HUBKA, V. *Principles of Engineering Design.* London: Butterworth Scientific, 1982.

KING, B. *Better Designs in Half the Time: Implementing Quality Function Deployment in America.* GOAL/QPC, Methuen, MA, 1989.

MIZUNO, S. and AKAO, Y. *QFD: The Customer-Driven Approach to Quality Planning and Deployment.* [Translated by Glenn Mazur]. Asian Productivity Organization, 1994.

MOEN, R., NOLAN, T., and PROVOST, L. P. *Improving Quality Through Planned Experimentation.* New York: McGraw-Hill, 1991.

NAKUI, S. "Comprehensive QFD." *Transactions of the Third Symposium on Quality Function Deployment.* Novi, MI, June 24–25, pp. 137–152, 1991.

PAHL, G. and BEITZ, W. *Engineering Design: A Systematic Approach.* New York: Springer-Verlag, 1996.

PUGH, S. *Total Design.* Reading, MA: Addison-Wesley, 1990.

SHILLITO, M. L. *Advanced QFD: Linking Technology to Market and Company Needs.* New York Wiley-Interscience, 1994.

ULMAN, D. G. *The Mechanical Design Process.* New York: McGraw-Hill, 1992.

VIDOSIC, J. P. *Elements of Engineering Design.* New York: The Ronald Press Co., 1969.

Developing Concepts

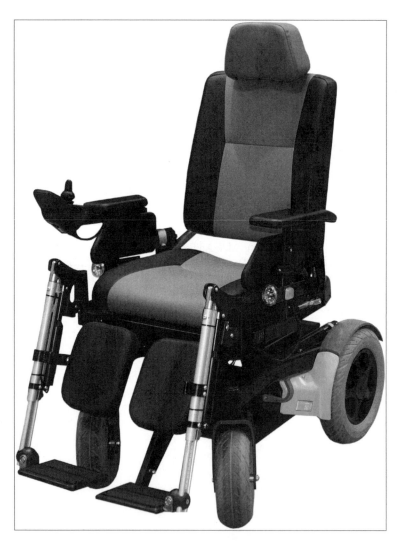

A design team is required to design a retrieval unit for wheelchairs to assist nurses performing walking activities with patients. The design team for this wheelchair retrieval unit develops a function analysis for the unit and then creates a morphological chart. (Daseaford/Shutterstock)

7.1 OBJECTIVES

By the end of this chapter, you should be able to

1.   Use systematic methods to generate conceptual designs.
2.   Generate a morphological chart.
3.   Improve your creative brainstorming techniques.

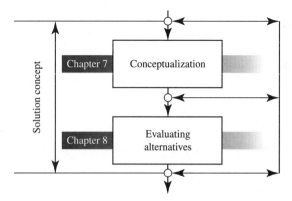

So far, we have developed methods to understand and define customers' needs as clearly as possible. We used objective trees to expand on the needs and determine the various levels of goals. We used functional trees to address the different functions that need to be carried out in a hierarchical manner to satisfy those needs. Then we drew up the specification chart based on what we needed to achieve and what we needed to do to achieve the objectives. The specifications that must be satisfied were classified as "must," and those that will simply enhance the product quality as "desired." However, specifications describe what needs to be achieved in a measurable quantity rather than as a qualitative desire expressed at the objective stage. The next step is to come up with more than one artifact that will satisfy the needs of the customer, taking into consideration the desires and wishes of the specifications. Now you have reached a stage where concepts need to be generated based on what you have accomplished in the previous steps of the design process. These concepts represent multiple solutions to the problem.

As was discussed in previous chapters, the goal of the design process is to develop artifacts that satisfy a need. Although the majority of designs can be considered redesigns or modifications of existing products, these modifications need to be new and unique. The systematic process of design that you have followed stresses not having your mind set on a specific solution. Thus, we have considered the process as producing several alternatives; later you will find that the process converges to one unique solution. Several design methodologies stress the concept of brainstorming as a model to generate several alternatives. In the presented design process, brainstorming plays an important role in generating alternatives for each function of the proposed solution. At this stage, brainstorming is important. However, you can't use brainstorming efficiently without understanding and going through a systematic process.

7.2 DEVELOPING WORKING STRUCTURES

Once the functions at various levels are known, it is important to find the principles at work for each of the functions and subfunctions. A working principle must be based on the physical effect needed to achieve the given function based on the flow of materials, energy, and information within each function. Each function may be achieved in a number of ways. One important tool that can be used effectively is a design catalog or morphological chart. Figure 7.1 shows the schematic of a morphological chart; each subfunction is shown in the left column and the many possible solutions in the corresponding rows. Figure 7.2 shows an example of this for a mechanical vegetable-collection system. The general idea is to identify as many means possible to achieve the same functional requirement.

Another frequently occurring design situation involves using the same working principle but changing the geometry or material combination (form) to enhance the functional requirement.

Possible Solution/ Options Sub Function		1	2	...	j	...	m
1	F_1	O_{11}	O_{12}		O_{1jj}	-	O_{1m}
2	F_2	O_{21}	O_{22}		O_{2jj}		O_{2m}
:		:	:		:		:
i	F_i	O_{i1}	O_{i2}		O_{ijj}		O_{im}
:							
n	F_a	O_{n1}	O_{n2}		O_{njj}		O_{nm}

F = Function of sub function
O = Options
ij = i,j solution/or option

Figure 7.1 Basic structure of a classification scheme with the subfunctions of an overall function and associated solutions.

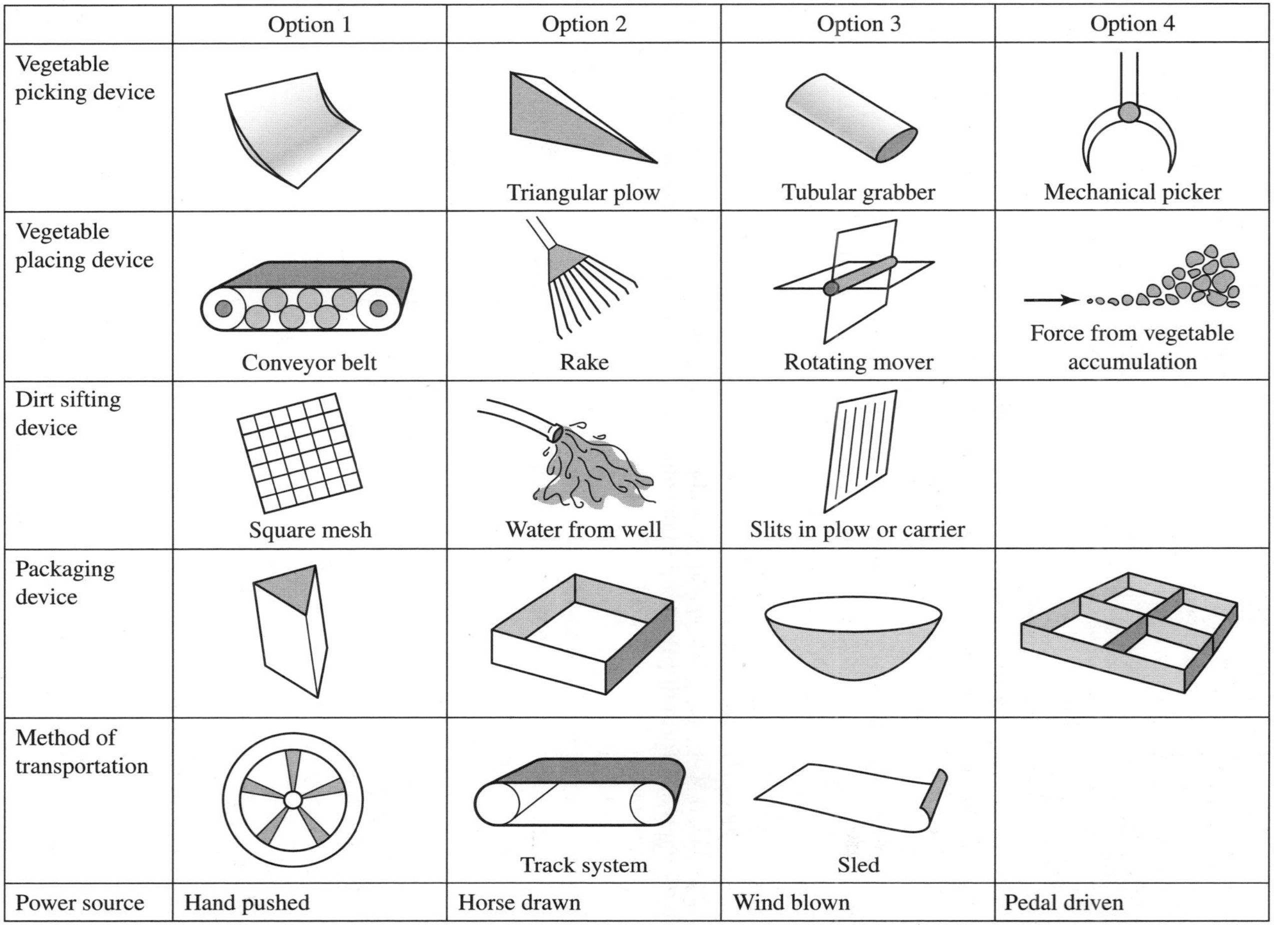

	Option 1	Option 2	Option 3	Option 4
Vegetable picking device	Conveyor belt	Triangular plow	Tubular grabber	Mechanical picker
Vegetable placing device		Rake	Rotating mover	Force from vegetable accumulation
Dirt sifting device	Square mesh	Water from well	Slits in plow or carrier	
Packaging device		Track system	Sled	
Method of transportation				
Power source	Hand pushed	Horse drawn	Wind blown	Pedal driven

Figure 7.2 Mechanical vegetable collection system.

175

7.3 STEPS TO DEVELOP CONCEPTS FROM FUNCTIONS

Step 1. *Develop concepts for each function:*
The goal of this first step is to generate as many concepts as possible for each of the functions identified in the decomposition. There are two activities here that are similar to each other. First, develop as many alternative options functions as possible for each function. Second, for each subfunction, the goal is to develop as many means of accomplishing the function as possible. In the previous example, vegetable placing could be achieved using a conveyor belt, a rake, a rotating mover, or simply by accumulating the vegetables. If there is a function for which there is only one conceptual idea, this function should be reexamined.

If there is a very limited idea, it may be due to the fact that the designer has made a fundamental assumption without realizing it. For example, one function typically encountered is to move in closely. If moving by hand is the only way, then the implicit assumption is that the part will be assembled manually. This may not be a bad assumption, except that it may not have a universal validity. If the function is to roll on the rail, then the general concept of motion has been specifically by indicating the specific movement along the rail. Designers may have a limited idea of other fields. It is generally important to keep the level of abstraction (details) at the same degree.

Step 2. *For each function or subfunction list all means or methods to be used:*
These secondary lists are the individual subsolutions that, when one is combined from each list, form the overall design solution. Each row represents all the possible subsolutions for the particular subfunction. The subsolutions can be expressed in rather general terms, but it is probably better if they can be identified as real devices or subcomponents. For example, when you want to lift something, you may like to use a ladder, screw, hydraulic piston, or rack/pinion. The list of solutions should contain not only the known components but also any new or unconventional methods or solutions for achieving the specific task.

Many brainstorming or lateral thinking concepts can be employed to arrive at the solution, and techniques to carry this out are discussed in the following section. Remember that many possibilities are necessary for each function. Also, when trying to identify artifacts for a given function, only that specific function or subfunction should be the criterion—not the overall function. Those eliminations will be taken care of at a different time in the design process.

Step 3. *Draw up the chart containing all of the possible subsolutions:*
This chart is called the morphological chart. The morphological chart is constructed from the functional list. At first, this is simply a grid of empty squares. The left-hand column lists each function identified in the functional diagram.

Across each row, it lists all alternative means (methods) for achieving the function. In the left-hand column, each method is entered in a square. As many methods as possible for each function should be entered. Once finished, the morphological chart contains the complete range of all of the different (theoretically possible) solution forms for the product. This complete range of solutions consists of the combinations made up by selecting one subsolution at a time from each row. The total number of combinations is rather large. The list of methods should be kept to about five.

Step 4.   *Identify feasible combinations:*
A list of concepts will be generated from the previous step for each function. The next step is to combine the individual concepts into complete conceptual designs. The method is to select one concept for each function and combine those selected into a single design. Some of these combinations, probably a small number, will be existing solutions; some new solutions; and some (usually a great number) impossible solutions.

Sometimes it is helpful to pick combinations based on 'themes' for each potential solution, for example, the economical one, the environmental one, the fancy one, etc. You should be careful though, of not limiting your choices by identifying inappropriate or limited themes. It is also essential to maintain an open mind at this point by assuming that all selected solutions should seem feasible. It is necessary to identify as many potential solutions as possible for further consideration and evaluation. Evaluating these solutions is the subject of Chapter 8. The next section discusses how brainstorming can help in generating ideas for morphological charts as well as potential solutions for the complete product.

7.4   BRAINSTORMING

The basic concept of brainstorming is to generate a large quantity of ideas. Research shows that the more ideas, the higher the quality of the desired product. A brainstorming session requires participants to be prepared to offer all ideas, including seemingly silly ideas. In many cases, silly ideas lead to genuine and creative ideas. Designers are encouraged to include nonengineers in the brainstorming sessions; often the best ideas come from people who do not have expert opinions on the area or subject at hand. Marketing and psychology experts also are often included in brainstorming design sessions.

In the systematic design process, you have used brainstorming to generate the objective tree and function analysis. Concept generation is an important step; thus, we focus on mechanisms and tools that can facilitate the improvement of the design output. Remember the iterative nature of the design process. If you have gathered information that you must update, you can do so at this stage.

7.4.1   Mechanism of Brainstorming Session

Participants must be prepared to be highly involved and active in the session. The topics of the brainstorming session need to be specific and stated clearly. Because of the energy level needed for such sessions, it is recommended to provide up to an hour for each topic/session. A set of rules has been developed for brainstorming sessions:

1. No criticism of ideas during the session. One way to prevent the criticism is to disallow discussion during the session. Present the ideas in brief and short format. One of the participants in the group will need to collect all ideas or use a sticker for each idea.

2. Wild, silly, and crazy ideas are welcomed. Such ideas may help others generate solid ideas or help maintain a fun and humorous environment.

3. Generate as many ideas as possible. One way to do so is to appoint a facilitator who will keep rotating the turns without allowing break time. Keep a competitive spirit alive. Some people produce more when competing with others in a group.

4. Adding to or improving presented ideas is welcomed. This will help maintain a stream of ideas.

7.4.2 Ideation

As an extension to the points raised previously, ideation is a process of generating ideas for a design solution by asking a set of structured questions. It relies on the fact that when there are several ideas, the chances for it to have a good one is higher than the chances are when the number of ideas are few. There are many techniques and different ways to structure questions. The following is one such example. These are a set of 15 classical questions aimed to guide thinking in particular directions in order to help generate new ideas for a product.

1. What is wrong with it?

   Make a list of all things that you feel are wrong with the present product, idea, or task.

2. How can I improve it?

   Forgetting feasibility, list all of the ways you would improve the present product, idea, or task.

3. What other uses does it have in its present form? What other uses are there if the idea is modified? Can it perform a function that was not originally intended?

4. Can it be modified?

   For example, change, trim, shape, description, weight, sound, form, contours, etc.

5. Can it be magnified?

   Make it larger, higher, longer, wider, heavier, or stronger.

6. Can it be Minified?

   Make it smaller, shorter, narrower, lighter, subtract something, or miniaturize.

7. Is there something similar you can adopt? What can be copied? Can it be associated with something else? is there something in stock or surplus that can be used?

8. What if I reverse it?

   Try a twist: opposites, upside down, turn around, rearrange, opposite pattern, opposite sequence, etc.

9. Can it have a new look?

   Change the color, form, or style: streamline, use a new package or new cover.

10. Can it be based on an old look?

   Copy a period, antique, parallel a previous winner, look for prestigious features. Trade on "They don't build them like that any more."

11. Can I rearrange it?

    Try a different order, interchange components, piece together differently, or change places.

12. Can it be substituted?

    What can take its place? Plastic for metal, metal for plastic, light instead of dark, round instead of square? What other process, principle, theory, or method can be used?

13. Can I combine ideas, principles, methods, groups, components, hardware, or issues.

14. Can I simplify?

    Make it easier, less work, easier to reach, disposable, simple to use, or quicker.

15. Can it be made safer?

    What devices, properties, controls, or sensors, can be added to prevent injury, accident, explosion?

You will be surprised by the number of new ideas you can generate using such a simple technique. Try it: Pick an existing product and go through the following questions thoroughly and allow sufficient time to consider each point. See how many new ideas you can come up with.

7.5 CREATIVITY

In the systematic design process, creativity is utilized in all steps. The need for group creative problem solving is utilized in constructing the objective tree, deducing information from market analysis, developing function analysis, structuring a house of quality, and developing concepts. As such, many design books provide a strong connection between design and creativity. Some research concludes that design is a creative process by nature. In this section, a definition of creativity in relation to design is introduced.

Ned Herrmann, author of *The Creative Brain*, defines creativity: "Creativity in its fullest sense involves both generating an idea and manifesting it—making something happen as a result. To strengthen creative ability. You need to apply the idea in some form that enables both the experience itself and your own reaction and others' to reinforce your performance. As you and others applaud your creative endeavors, you are likely to become more creative." Lumsdaine defines creativity: "Playing with imagination and possibilities, leading to new and meaningful connections and outcomes while interacting with ideas, people and environment." Kemper has the following to say about creativity: "The word creativity is a rather curious state: Nearly everyone can recognize creativity instantly, yet no one can define the word in a fully acceptable manner. A companion word, invention, causes even more trouble. Every layperson is instantly confident of the word invention and knows exactly what it means; such a person can think of countless inventions: The electric light, the safety pin, even the atomic bomb. It is only Patent Office examiners and United States Supreme Court justices who believe there is a problem in defining what an invention is."

In these definitions, we see that creativity is associated with generating ideas. We need to ask if creativity is an individual quality or can be associated with a group or team.

Furthermore, the more important question is if creativity is natural or can be taught. If creativity is natural, then creativity can be categorized in each individual as either present or absent. However, research has found that there is a degree of creativity in each individual that depends on the cognitive style, personality, and the creative outcome psychologists suggest that the level of creativity can be determined either statistically or assessed by experts.

Research has found that groups perform better on creative problem-solving tasks. It has been said that two heads are better than one, which also may be applied to creativity. Brainstorming is the best-known and most widely used technique for idea generation in groups. However, for brainstorming sessions to outperform individuals in generating ideas, participants must

1. have some social relationships (e.g., as students you identified with each other the first day you walk into the classroom).
2. have used some of the idea generated (you have done this through the objective tree and function tree);
3. have some technical experience pertinent to the problem (although not as important but preferable); and
4. have worked some tasks interdependently.

7.5.1 How to Increase Your Level of Creativity

The author John Steinbeck once said, "Ideas are like rabbits. You get a couple and learn how to handle them, and pretty soon you have a dozen." The following points will help you become more creative.

Know Your Thinking Style Since early history, humans have been trying to understand how the brain works. Herrmann developed a metaphorical model of the brain that consists of four quadrants. This is covered in more detail in Chapter 2 of this book. In team work, you may encounter different thinking styles. A good team is one that represents a full brain. You can train yourself to have all quadrants function at the same power or increase activities toward utilizing more of a specific quadrant.

1. Identify the weakness.
2. Attack problems that require the utilization of a weak quadrant.

Use Visual Imagery Einstein asserted that imagination is more important than knowledge, for knowledge is finite whereas imagination is infinite. Bernard Shaw said, "You see things and say why? But I dream things that never were and I say why not?" The inner imagery of the mind's eye has played a central role in the thought processes of many creative individuals. Albert Einstein ascribed his thinking ability to imagining abstract objects.

Kekule discovered the all-important structure of the benzene ring in a dream. Most visual thinkers clarify and develop their thinking with sketches. Drawing not only helps them to bring vague inner images into focus; it also provides a record of the advancing thought stream. Sketches expand the ability to hold and compare more information than is possible from simple recall from the brain. Sketches are usually extensions of thought, and they provide a conference table to talk to oneself. Distinguishing between the state of sleep and awake and between visions and perceptions has been studied since ancient history. According to researchers in visual imagery, there are three states of visual imagery:

a. Dharana, in which you simply focus on a given place or object.
b. Dhydana, in which you strengthen the focus of attention.
c. Samadhi, in which you experience a union fusion with the object and cannot distinguish between self and object.

As students, you have been practicing memorization since kindergarten; visualization will help enhance your memory as each image stores much more information than words. The recollection of these images will enhance your idea generation.

Reframing Einstein asserted that problems cannot be solved by thinking within the framework in which the problems were created. Reframing involves taking problems out of their frame and seeing them in a different context. It allows consideration of potentially valuable ideas outside current frames. The most common habits that limit the ability to change mental frames are as follows:

a. Pursuit of perfection: Many people are under the false impression that working long and hard is sufficient to develop the perfect answer to the problem.
b. Fear of failure: Resistance to change because others may perceive you as incompetent.
c. Delusion of already knowing the answer: Once you find an answer, you do not look for others. Why look for another answer if this one works?
d. Terminal seriousness: Many people are under the illusion that humor and serious idea generation do not mix.

Humor It was said that "Men of humor are always in some degree men of genius." The dictionary definition of humor is "The mental faculty of discovering, expressing or appreciating ludicrous or absurdly incongruous elements in ideas, situation, happenings or acts." If you were to take the words *ludicrous* or *absurdly* out of the definition of humor, you would have a definition of creativity.

Information Gathering In market analysis (Chapter 3), you were introduced to the power of gathering information on the problem statement. Gathering information will also enhance creativity; it will allow you to view ideas generated by other creative minds (written brainstorming). Patents can serve as an excellent source of ideas. However, it is difficult to identify the specific patent that contains the idea that you are looking for.

There are two main types of patents: utility patents and design patents. Utility patents deal with how the idea works for a specific function. Design patents only cover the look or form of the idea. Hence, utility patents are very useful, since they cover how the device works not how it looks. Since there are about 5 million utility patents, it is important to use some strategies to hone in on the one that you may be able to use. Use a Web search, such as that provided by the Patent Office. Key word searches (as well as patent numbers, inventors, classes, or subclasses) are available.

The next place to find information is in reference books and trade journals in the relevant area. Another information source is to consult experts. If designing in a new domain, you have two choices for gaining the knowledge sufficient for generating concepts. You either find someone with expertise, or you spend time gaining experience on your own. Experts are those who work long and hard in a domain, performing many calculations and experiments themselves to find out what works and what does not. If you cannot find or afford an expert, then the next best source is the manufacturer's catalog or manufacturer's representative.

The mechanism for developing concepts is discussed in the following few sections.

7.6   DEVELOPING CONCEPTS—SAMPLES

7.6.1   Mechanical Vent

Consider the case involving a design to automate the opening and closing of air conditioning vents in a centralized location in the house (see Chapter 3). Design teams have developed a function analysis wherein the functions are

1. Select vent
2. Send signal
3. Receive signal
4. Convert signal
5. Open/close vent

Figure 7.3 shows a morphological chart for this problem. The five functions are entered in the left-hand column of the chart. Each row represents a specific function.

For example, row 5 is the function "open/close vent," and we need to find means to achieve the function of opening/closing the vent. The chart shows five different means or methods to achieve the function using gears, belt, electric field, cable, or impact plate. The methods are entered using words and diagrams in each of the five squares following "open/close vent." The chart in Figure 7.3 thus contains the five functions and the different (5, 5, 5, 6, 5) ways of achieving each function. Thus, theoretically, there are 3750 possible different open/close vent machines, combining the various means. However, many of them are not viable. At least five or six of such combinations will be identified for building the system.

7.6.2   Wheelchair Retrieval Unit

A design team is required to design a retrieval unit for wheelchairs to assist nurses performing walking activities with patients. In most cases, a single nurse is in charge of assisting patients during walking exercise. However, the nurse cannot assist the patient and drag the wheelchair. The design team developed a function analysis for the unit as follows:

1. Align wheelchair to patient and nurse
2. Move wheelchair
3. Steer wheelchair
4. Stop wheelchair

Figure 7.4 on page 180 shows the morphological chart for this unit. It consists of six different options for each of the four functions, thus you will have 64 possible solutions.

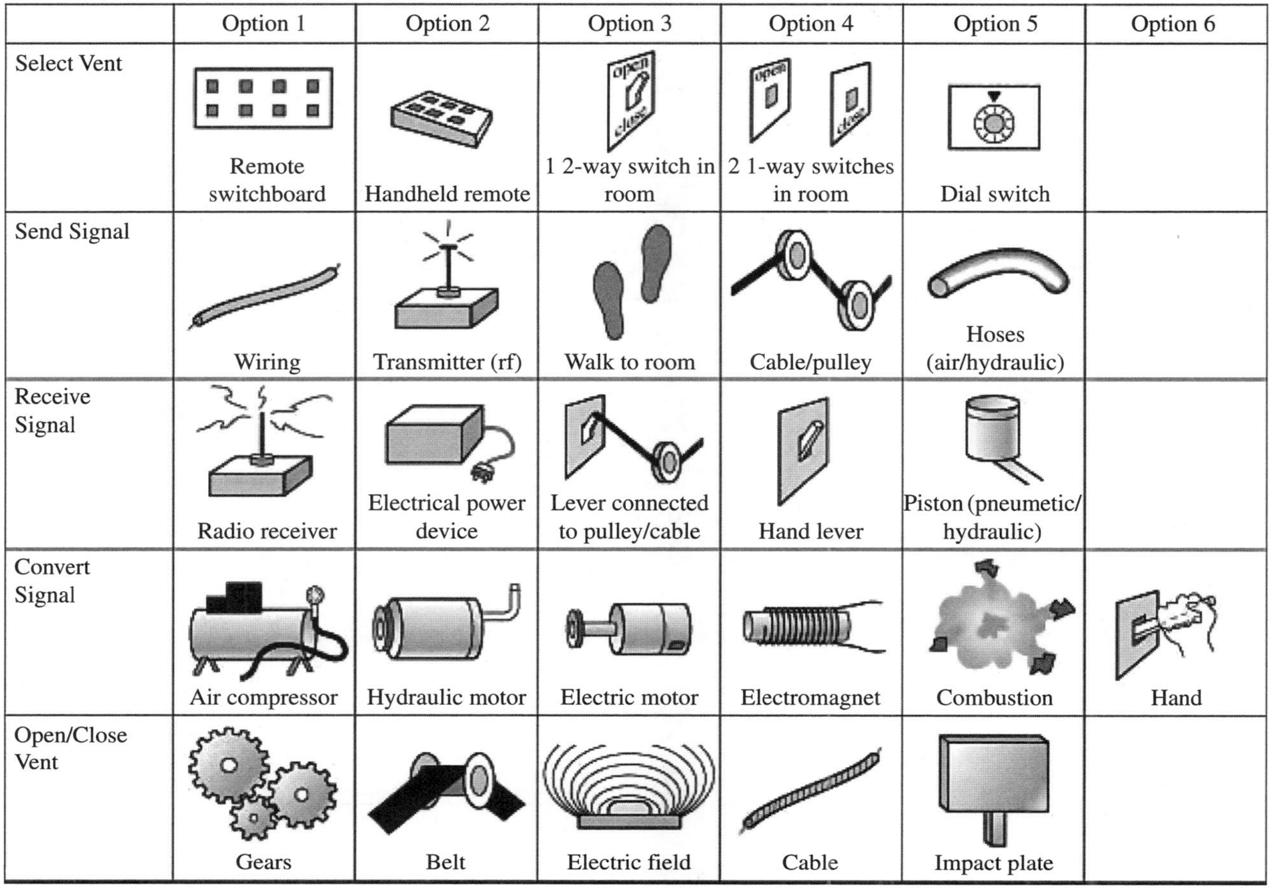

	Option 1	Option 2	Option 3	Option 4	Option 5	Option 6
Select Vent	Remote switchboard	Handheld remote	1 2-way switch in room	2 1-way switches in room	Dial switch	
Send Signal	Wiring	Transmitter (rf)	Walk to room	Cable/pulley	Hoses (air/hydraulic)	
Receive Signal	Radio receiver	Electrical power device	Lever connected to pulley/cable	Hand lever	Piston (pneumetic/hydraulic)	
Convert Signal	Air compressor	Hydraulic motor	Electric motor	Electromagnet	Combustion	Hand
Open/Close Vent	Gears	Belt	Electric field	Cable	Impact plate	

Figure 7.3 Morphological chart of mechanical vent machine.

183

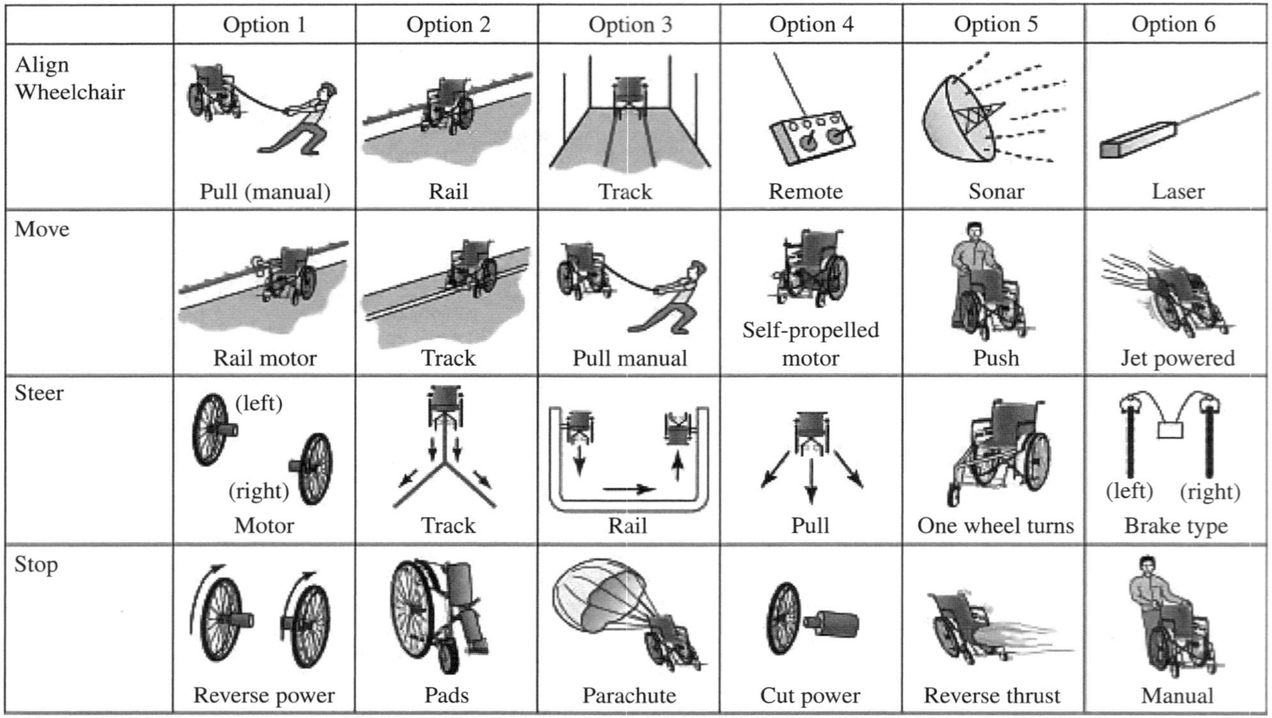

	Option 1	Option 2	Option 3	Option 4	Option 5	Option 6
Align Wheelchair	Pull (manual)	Rail	Track	Remote	Sonar	Laser
Move	Rail motor	Track	Pull manual	Self-propelled motor	Push	Jet powered
Steer	(left) (right) Motor	Track	Rail	Pull	One wheel turns	(left) (right) Brake type
Stop	Reverse power	Pads	Parachute	Cut power	Reverse thrust	Manual

Figure 7.4 Wheelchair retrieval unit options.

Again, five or six different viable solutions will be selected out of these solutions. For example, one possible combination is rail, track, rail, reverse power. (i.e., 2,2,3,1).

7.6.3 Automatic Can Crusher

The design team lists all the functions that must be accomplished. Then team members generate a matrix that shows the functions in the right column, and they point out different ways to achieve these functions. Creativity should be exercised during this activity. The different possibilities may be listed in text or in sketches. Both methods are exercised in this situation. The available power sources are

- Hydraulic
- Magnetic
- Gravity
- Thermal
- Sound
- Pneumatic
- Solar
- Electric
- Combustion

Several designs can be generated from the morphological chart shown in Figure 7.5 on the next page.

7.7 PROBLEMS

7.7.1 Team Activities

1. Develop a morphological chart for the following statement. If you drive in the state of Florida, you may notice some traffic congestion due to the trees being trimmed on state roads and freeways. Assume that the Florida Department of Transportation is the client. Usually, branches 6 inches or less in diameter are trimmed. The allowable horizontal distance from the edge of the road to the branches is 6 ft. The material removed from the trees must be collected and removed from the roadside. To reduce the cost of trimming, a maximum of only two workers can be assigned for each machine. The overall cost, which includes equipment, labor, etc., needs to be reduced by at least 25% from the present cost. The state claims that the demand for your machines will follow the price reduction (i.e., if you are able to reduce the cost by 40%, the demand will increase by 40%). Allowable working hours depend on daylight and weather conditions.

2. Compare (a) the systematic design process, in which functions are created—based on these functions—a morphological chart is created, and thus, a large number of possible solutions; with (b) a design process, in which the problem is identified and then a brainstorming session is started to generate different

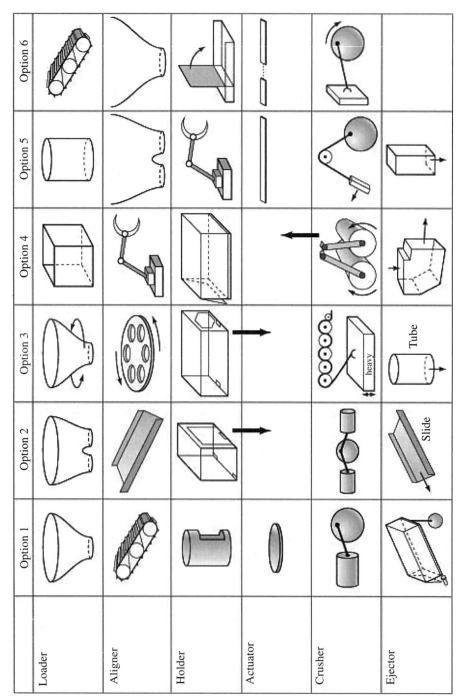

Figure 7.5 Morphological chart of automatic can crusher.

186

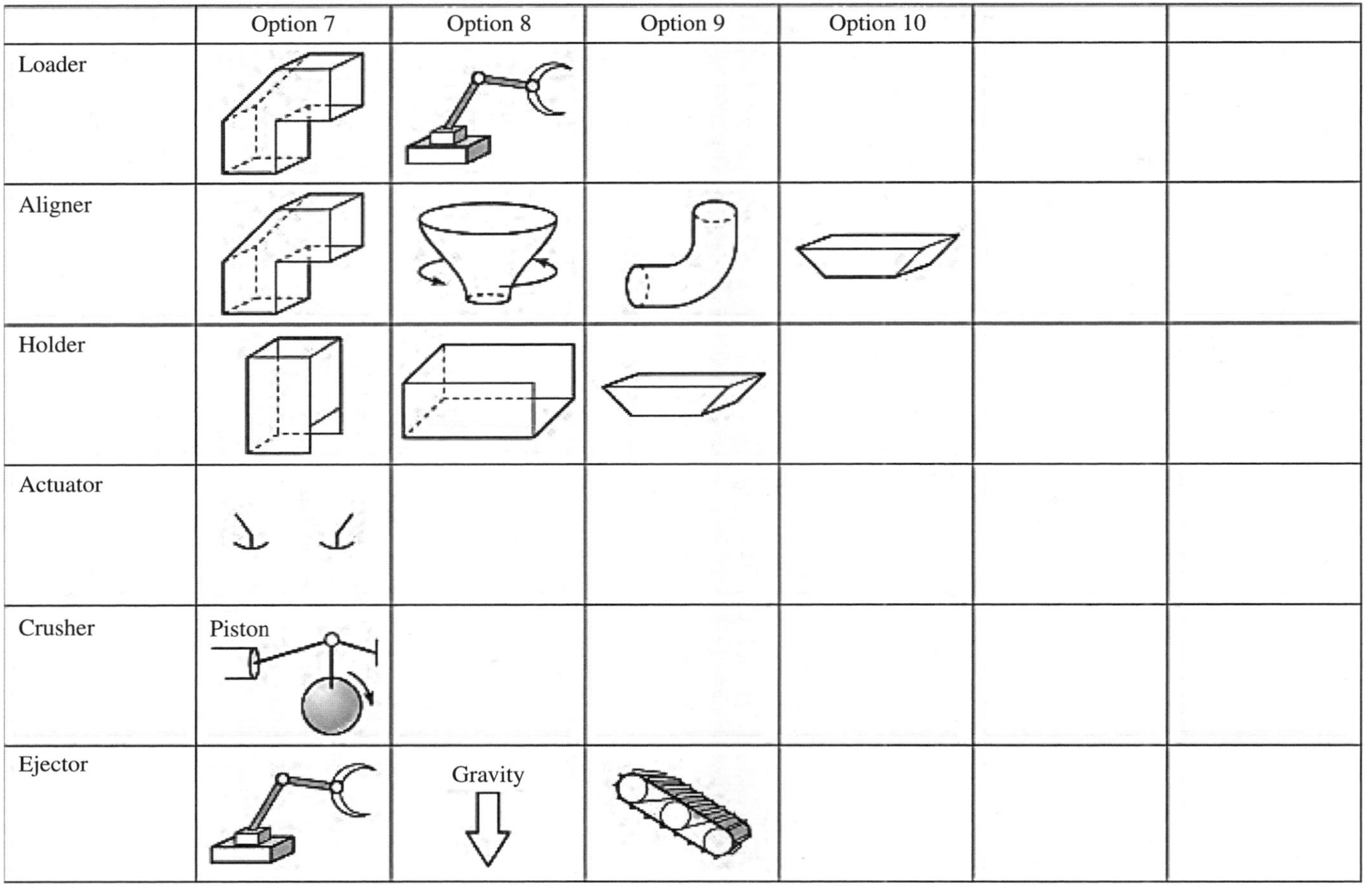

	Option 7	Option 8	Option 9	Option 10		
Loader						
Aligner						
Holder						
Actuator						
Crusher	Piston					
Ejector		Gravity				

Figure 7.5 Morphological chart of automatic can crusher (continued).

187

solutions. Which of these methods would you utilize. Why? In the event that a certain aspect of the design needs modification, which of the two methods would be easier to apply toward that modification. Why?

3. Compare and contrast group creativity and individual creativity.
4. Discuss the statement that design is a social activity. How would that statement fit in the systematic process?

7.7.2   Individual Activities

1. What is the difference between creativity and innovation?
2. What is a morphological chart?
3. Without a creative objective tree, you would not have a creative morphological chart. Discuss.
4. How would you increase your creativity level? Categorize yourself (analytical, organizational, social, intuitive). You can perform the survey in the lab exercise in Chapter 1 to find out (see Lab 1).
5. What is the Herrmann model of the brain?

7.8   Selected Bibliography

AMABILE, T. M. and CONTI, R. "Changes in the Work Environment for Creativity During Downsizing." *Academy of Management Journal*, Vol. 42, pp. 630–640, 1999.

CRANDALL, R. *Break-Out Creativity*. Corte Madera, CA: Select Press, 1998.

CROSS, N. *Engineering Design Methods: Strategies for Product Design*. New York: Wiley, 1994.

DHILLON, B. S. *Engineering Design: A Modern Approach*. Toronto: Irwin, 1995.

DUNNETTE, M. D., CAMPBELL, J., and JAASTAD, K. "The Effect of Group Participation on Brainstorming Effectiveness for Two Industrial Samples." *Journal of Applied Psychology*, Vol. 47, pp. 30–37, 1963.

DYM, C. L. *Engineering Design: A Synthesis of Views*. Cambridge, UK: Cambridge University Press, 1994.

GALLUPE, R. B., BASTIANUTTI, L. M., and COOPER, W. H. "Unblocking BrainStorms." *Journal of Applied Psychology*, Vol. 76, pp. 137–142, 1991.

GARDNER, H. "Creative Lives and Creative Works: A Synthetic Approach." *In The Nature of Creativity*, J. R. Sternberg (editor), pp. 298–321. Cambridge, UK: Cambridge University Press, 1988.

GOLDENBERG, J. and MAZURSKY, D. "First We Throw Dust in the Air Then We Claim We Can't See: Navigating in the Creativity Storm." *Creativity and Innovation Management*, Vol. 9, pp. 131–143, 2000.

HENNESSEY, B. A. and AMABILE, T. M. "The Conditions of Creativity." In *The Nature of Creativity*, J. R. Sternberg (editor), pp. 11–38. Cambridge, UK: Cambridge University Press, 1988.

HERRMANN, N. *The Creative Brain*. Lake Lura, NC: Brain Books, 1990.

HILL, H. W. "Group Versus Individual Performance: Are N_2 Heads Better Than One?" *Psychological Bulletin*, Vol. 91, pp. 517–539, 1982.

ISAKSEN, S. G. and TREFFINGER, D. J. *Creative Problem Solving: The Basic Course*. Buffalo, NY: Bearly Limited, 1985.

KEMPER, J. D. *Engineers and Their Profession*. New York: Oxford University Press, 1990.

KOLB, J. "Leadership of Creative Teams." *Journal of Creative Behavior*, Vol. 26, pp. 1–9, 1992.

LUMSDAINE, E. M., LUMSDAINE, M., and SHELNUTT, J. W. *Creative Problem Solving and Engineering Design*. New York: McGraw-Hill, 1999.

MCCABE, M. P. "Influence of Creativity on Academic Performance." *Journal of Creative Behavior*, Vol. 25, p. 2, 1991.

MCKIM, R. H. *Experiences in Visual Thinking*. Boston, MA: PWS Publishing Co., 1980.

MUMFORD, M. D. and GUSTAFSON, S. B. "Creativity Syndrome: Integration, Application and Innovation." *Psychological Bulletin*, Vol. 103, pp. 27–43, 1988.

NICKERSON, R. S. "Enhancing Creativity." In *The Nature of Creativity*, J. R. Sternberg (editor), pp. 392–430. Cambridge UK: Cambridge, University Press, 1988.

PAHL, G. and BEITZ, W. *Engineering Design: A Systematic Approach*. Springer-Verlag, 1996.

ULMAN, D. G. *The Mechanical Design Process*. New York: McGraw-Hill, 1992.

ULRICH, K. T. and EPPINGER, S. D. *Product Design and Development*. New York: McGraw-Hill, 1995.

WEISBERG, R. W. *Creativity: Genius and Other Myths*. New York: Freeman, 1986.

WHITE, D. "Stimulating Innovative Thinking." *Research-Technology Management*, Vol. 39, pp. 31–35, 1996.

CHAPTER •8

Concepts Evaluation

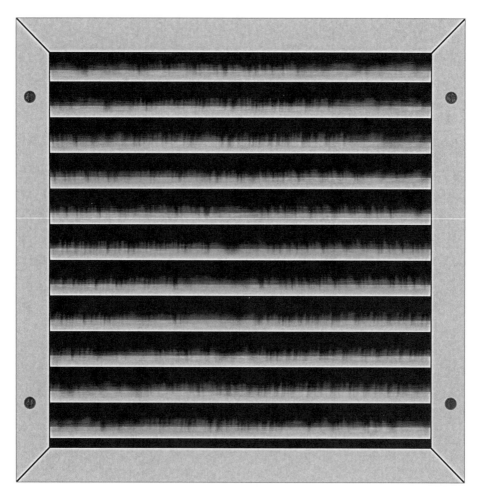

An engineer or design team will determine a product's functional alternatives from its morphological chart. This will assist the design team in finding the best design for the client's need. Having chosen a set of alternatives, the design team can start sketching them as devices. These vents were the result of numerous design sketches. (Ivan Cholakov Gostock-dot-net/Shutterstock)

8.1 OBJECTIVES

By the end of this chapter, you should be able to
1. Use different methods to evaluate the different concepts that were generated in previous design step.
2. Select a design alternative for further development.

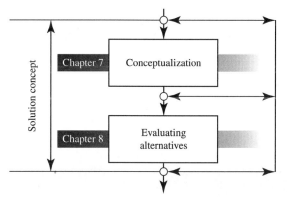

In the previous chapter, we developed a technique for generating different concepts that satisfy the need statement. As was demonstrated, the number of possibilities that may satisfy the need is a function of the number of entries (different ways) that fulfill a function. The larger the number of entries, the larger the number of concepts generated. Obviously, not all combinations through the morphological chart will lead to a feasible product. The goal of this chapter is to provide a methodology for evaluating the different devices. By the end of the evaluation process, a product that has the highest potential for becoming the quality product will be identified through the many concepts that were generated by the morphological chart.

The difficulty in concept evaluation resides in that the design team members must choose a concept to develop based on limited knowledge and data. Concepts at this point are still abstract, have little details, and cannot be measured. The essential question then becomes, "Should time be spent refining and adding minute details to all concepts and then measuring their outputs?" or "Would developing a mechanism to point out one concept that will most likely become the quality product prove to be a more efficient option?" The screening of concepts can be accomplished by using the following levels.

1. The concepts are directly measured with respect to a fixed set of limits (specifications). If a concept doesn't satisfy the specification requirement, it will be dropped. Although certain functional presentations are on the morphological chart, they will be eliminated from assembly if they do not meet the specification criteria established earlier. For example, using sonar to align the wheelchair may be eliminated in favor of another functional presentation that conforms to the projects cost limitations.

2. The design team uses their experience (engineering sense) to eliminate some concepts because the technology is not yet available to facilitate the concept. Or, based on their judgment, the product is not feasible. Why use a parachute to stop the wheelchair if another function that is more sound (from an engineering standpoint) could perform the same function?

3. The concepts are evaluated with respect to each other using measures defined by the criteria (specifications/wishes). This screening method was developed by Stuart Pugh in the 1980s and is often called Pugh concept selection. The purpose of this stage is to narrow the number of concepts quickly. This evaluation matrix will be presented in more detail in this chapter.

8.2   SKETCH ASSEMBLY OF ALTERNATIVES

Using the different functional alternatives presented in the morphological chart, your engineering sense, and a specification table, choose a set of five to six alternatives.
At this stage, you have started the convergent part of the design process that will lead toward a best design for the need at hand.

Having chosen the set of alternatives, you need to sketch them as devices, not as a set of functions. The conceptual sketches are not detailed. The sketches serve as a first presentation of the prospective design. Also, sketching reveals information about details that may be needed later, which will also help in evaluating the concepts. Assembly of the functional mechanisms will help the visual conceptualization of the alternatives and will provide a presentation tool to the customer.

Figure 8.1 shows six different alternatives generated for the mechanical vent. Sketches are the artistic expression of the design. Visualizations of the creative ideas are presented in sketches. Students need to master the craft of sketching.

8.3   EVALUATING CONCEPTUAL ALTERNATIVES

Now that you have established several concepts, it is time to begin the process in selecting the best one. Two similar methods are presented next and essentially follow the same principle. They are *Pugh's evaluation matrix* and *the decision matrix*. Both methods are based on comparing alternatives with the list in the specification table. An alternative should meet the customer demands; otherwise it will be dropped in the initial screening. Also, the concept should seem feasible to the engineering group, or it will be dropped by the second stage of screening. The concepts that pass the initial two stages need to be evaluated with respect to each other, using fixed criteria (specification table). The Pugh evaluation method tests the completeness and understanding of requirements, rapidly identifies the strongest alternative, and helps foster new alternatives.

Both methods are most effective if each member of the design team performs the agreed upon method independently, and the individual results are then compared. The results of the

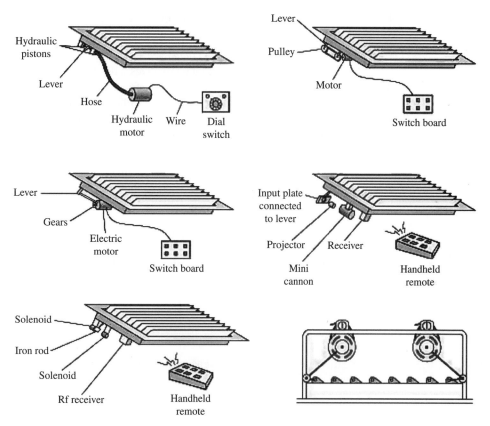

Figure 8.1 Conceptual sketches of the mechanical vent.

comparison lead to a repetition of the technique with the iteration continued until the team is satisfied with the results. The Pugh evaluation method is performed using the following steps.

Step 1. *Choose the comparison criteria:*
If all alternatives fulfill the demands on the same level, then the criteria should be listed in the specification table or the design criteria. Remember, the design criteria were organized in an ordered fashion before they were used in the house of quality. The order is based on the level of importance according to the design team. Different weighing schemes could be used in the ordering of wishes:

    a. Absolute factor, where each wish is evaluated individually on a scale from 0 to 10. Other wishes will not interfere with the weigh factor used.

    b. Relative scale, where the sum of the wishes is assigned a scale from 0 to 100. Each wish is assigned a weighing factor that corresponds to its importance relative to the other wishes. However, when the weighing factors are added, they should equal 100.

    In this book, the choice of the evaluation scheme is given to the design team. The weighting factor method is based on the engineering team's assessment of how important the attribute may be to the final product. If the alternatives contain differences in the levels at which demands may be fulfilled, then the demands can be entered in the comparison criteria.

Step 2. *Select the alternatives to be compared:*
From the morphological chart, different alternatives will be generated. Some of these alternatives will be dropped because they do not satisfy the customer demands or are not feasible. The rest of the alternatives are possible candidates. However, by using the initial screening stages, a few alternatives will be left for the final stage. Not all feasible alternatives will be allowed to enter the final stage. Up *to six will be allowed.*

Step 3. *Generate scores:*
The above points are common for either Pugh's Evaluation Matrix or the Decision Matrix. The following subsections describe how each method differs in the scoring.

8.3.1 Pugh's Evaluation Matrix

After careful consideration, the design team chooses a concept to become the benchmark or datum against which all other concepts are rated. The reference is generally either

- An industry standard which can be commercially available product or an earlier generation of the product.
- An obvious solution to the problem.
- The most favorable (measured according to a vote by the design team) concepts from the alternatives under consideration.
- A combination of subsystems which have been combined to represent the best features of different products.

For each comparison, the concept being evaluated is judged to be better than, about the same as, or worse than the datum. If it is better than the datum, the concept is given a positive [+] score. If it is judged to be about the same as the datum, the concept is given zero [0]. If the concept does not meet the criterion as well as the datum does, it is given a negative [−].

After a concept is compared with the datum for each criterion, four scores are generated: the number of plus scores, the number of minus scores, the overall total, and the weighted total. The overall total is the difference between the number of plus scores and the number of minus scores. The weighted total is the sum of each score multiplied by the weighing factor. The scores can be interpreted in a number of ways.

a. If a concept or a group of similar concepts has a good overall total score or a high positive total score, it is important to notice what strengths it exhibits. In other words, notice which criteria it meets better than the datum. Likewise, the grouping of negative scores will show which requirements are difficult to meet.

b. If most concepts get the same score on a certain criterion, examine that criterion closely. It may be necessary to develop more knowledge in the area of the criterion to generate better concepts. It may be that the criterion is ambiguous, or it may be interpreted differently by different members of the design team.

c. To learn even more, redo the comparison with the highest scoring concept used as the new datum. This iteration should be redone until what is clearly the best concept or concepts emerge. *This process becomes a must when the datum is one of the alternatives under consideration.*

After each team member has completed this procedure, the entire team should compare their individual results. Then a team evaluation needs to be conducted.

8.3.2 Decision Matrix

This method utilizes a more numerical approach. Here, the conceptual ideas are rated against criteria the designer decides as relevant to the task. Conceptual ideas go on the rows and the design criteria on the columns. The designer decides on importance ratings for each design criteria. For example, let us assume a simple product has only three criteria: *long, fast*, and *strong*, each with the following importance weighting:

- *Long*—35%
- *Fast* is very important—50%
- *Strong* is the least important—15%

Remember, all criteria weightings must add up to 100%. In the actual matrix, these would be scaled down to a total of 1.0, therefore long would be 0.35, fast would be 0.5, and strong would be 0.15. These numbers are referred to as the *weighting factors* or W.F.

For each concept, the designer tries to see how well it achieves each design criteria on a rating from 1 to 10—10 being the best. These numbers are referred to as the *rating factors* or R.F.

This example studies several concepts for a yard leaf remover—to remove leaves that have fallen from trees in a garden. Here are the design criteria the designer decides on.

Use of standard parts	8%
Safety	12%
Simplicity and maintenance	10%
Durability	10%
Public acceptance	18%
Reliability	20%
Performance	15%
Cost to develop	3%
Cost to buyer	4%

These values are scaled down to add up to 1.0 and are referred to as the W.F. (e.g., the weighting factor for the use of standard parts is 0.08 and for safety is 0.12). The designer then develops four conceptual alternatives to satisfy the product's requirements following a similar process as described throughout this book. The four concepts are displayed in Figure 8.2 on the next page. Each concept is then rated against each design criteria on a scale from 1 to 10 (the rating factor), and this is placed in the top-left triangle of each corresponding cell within

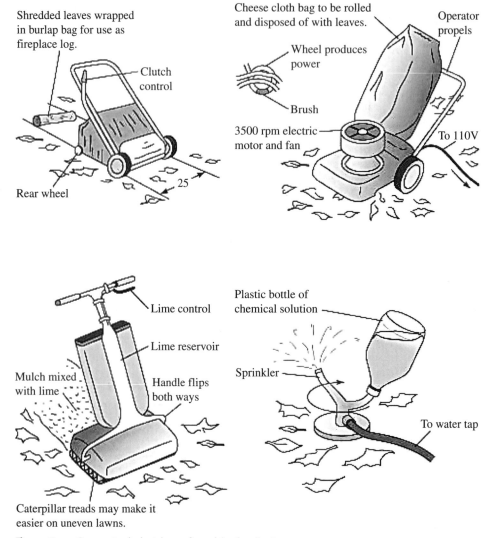

Shredded leaves wrapped
in burlap bag for use as
fireplace log.

Clutch
control

Rear wheel

25

Cheese cloth bag to be rolled
and disposed of with leaves.

Operator
propels

Wheel produces
power

Brush

3500 rpm electric
motor and fan

To 110V

Lime control

Lime reservoir

Mulch mixed
with lime

Handle flips
both ways

Caterpillar treads may make it
easier on uneven lawns.

Plastic bottle of
chemical solution

Sprinkler

To water tap

Figure 8.2   Conceptual sketches of yard leaf collector.

the decision matrix (Figure 8.3). Therefore, the *rating factor* for the leaf bailer concept is 3 for 'Use of Standard Parts' and is 5 for 'Safety' and so on, as can be seen in Figure 8.3.

The bottom-right triangle of each cell is calculated as W.F. × R.F. and presents the *weighted rating factor* of each conceptual alternative with respect to the individual design criteria. E.g., the weighted rating factor for the leaf bailer for 'Use of Standard Part' would be calculated as

$$\text{R.W.F.} = \text{R.F.} \times \text{W.F.} = 3 \times 0.08 = 0.24$$

For the same concept with respect to safety,

$$\text{R.W.F.} = \text{R.F.} \times \text{W.F.} = 5 \times 0.12 = 0.60$$

The weighted rating factors for each conceptual alternative are then added, and the final sums for each alternative are compared. The concept with the highest rating is the one that most

Design criteria / Weighting factor	Use of standard parts	Safe	Simplicity and maintenance	Dura-bility	Public acceptance	Reliability	Cost to develop	Cost to buyer	Perform-ance	Sum
Alternatives	0.08	0.12	0.10	0.10	0.18	0.20	0.03	0.04	0.14	1.0
A) Leaf bailer	3 / 0.24	5 / 0.60	2 / 0.20	4 / 0.40	9 / 1.62	6 / 1.20	1 / 0.03	1 / 0.04	3 / 0.45	4.78
B) Vacuum collector	9 / 0.72	10 / 1.20	10 / 1.00	8 / 0.80	6 / 1.08	7 / 1.40	10 / 0.30	10 / 0.40	8 / 1.24	8.14
C) Shredder	5 / 0.40	6 / 0.72	7 / 0.70	7 / 0.70	8 / 1.44	6 / 1.20	3 / 0.09	4 / 0.16	5 / 0.75	6.16
D) Chemical decomposer	8 / 0.64	10 / 1.20	9 / 0.90	8 / 0.80	9 / 1.62	7 / 1.40	2 / 0.06	8 / 0.32	8 / 1.24	8.18

Weighting factor (W.F.) = Measure of relative importance (0 to 1.0 Σ = 1.0)

Rating factor (R.F.) = Measured value of alternatives against design criteria (0 to 10)

Figure 8.3   Decision matrix.

closely satisfies the set design criteria. In the example given in Figure 8.3, both the chemical decomposer and vacuum collector have similar top ratings. In this case, common sense is also required to see which design is more suitable. It is also possible to try to incorporate some of the strong points of one design into the other if this is feasible in order to make it a better design.

8.4 CONCEPTS EVALUATION: MACHINE SHOP KIT

A team of design students was asked to design a steam-powered machine shop kit that can be used to (1) develop hands-on skills in using machine shop tools for freshman engineering students, and (2) demonstrate the conversion of thermal energy into work, thus becoming a demonstration tool for an introduction to thermal science class. Stirling engine kits are being used in many engineering schools. The new kit must compete with the Stirling engine kit in its educational value and its cost. A Stirling engine kit is a kit containing the disassembled parts that compose a Stirling engine; a few of the parts require students to use the equipment in the workshop.

Figure 8.4 shows the functional analysis that the design team developed. The energy track of the functions consists of receive energy, channel energy, store energy, transmit

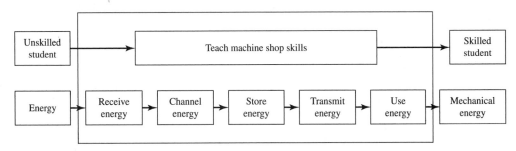

Figure 8.4   Function analysis of machine shop kit.

energy, and utilize energy. Figure 8.5 shows the house of quality of the problem. Figure 8.6 shows the morphological chart developed by the design team. Six different conceptual designs were developed in this exercise. Figures 8.7 through 8.12 (pp. 196–198) show these different concepts.

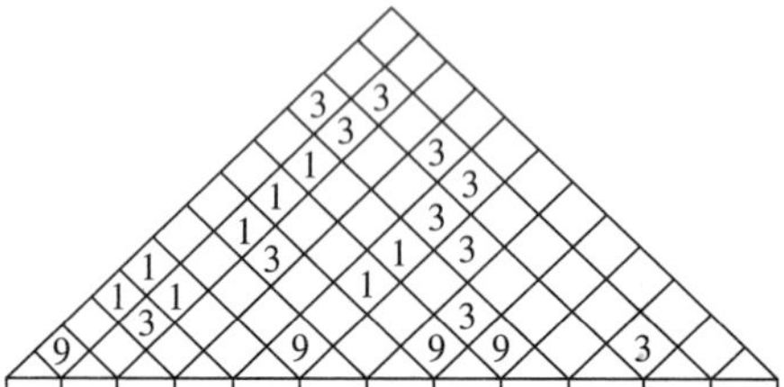

	Assembly time (<20 hrs)	# of parts (10–15)	Pollution factor (<9 ppm)	Injury probability (<0.1%)	Retail price (<$135)	Replace parts cost (<5–10% total cost)	Efficient (>40%)	Vibrations (<2/sec)	Noise (<60 dB)	Dimensions (1'×6"×6")	Weight (<20 lbs)	Users find visually pleasing (75%)
Assembly												
Easy to assemble	9	3		1								
Easy to disassemble	9	3		1								
Moderate assembly time	9	3										
Interesting to build	1											
Not too many parts	9	9			3	1		3	1	1	3	
Safety												
Low pollution			9				1					
No flying debris				9								
No sharp edges				9								
Costs												
Retails for less than the competition					9	1						
Low replacement part costs					3	9						
Inexpensive materials					9	3					3	1
Performance												
Convert energy efficiently		3			1		9	3				
Low vibration	1						3	9	3		1	
Runs off small amount of energy							3	1				
Low noise							1	3	9			
Physical requirements												
Portable										9	9	
Strong material					3			1			3	
Corrosion proof					1							3
Lightweight		3		1				1		3	9	
Visually appealing												9

Figure 8.5   House of quality for machine shop kit.

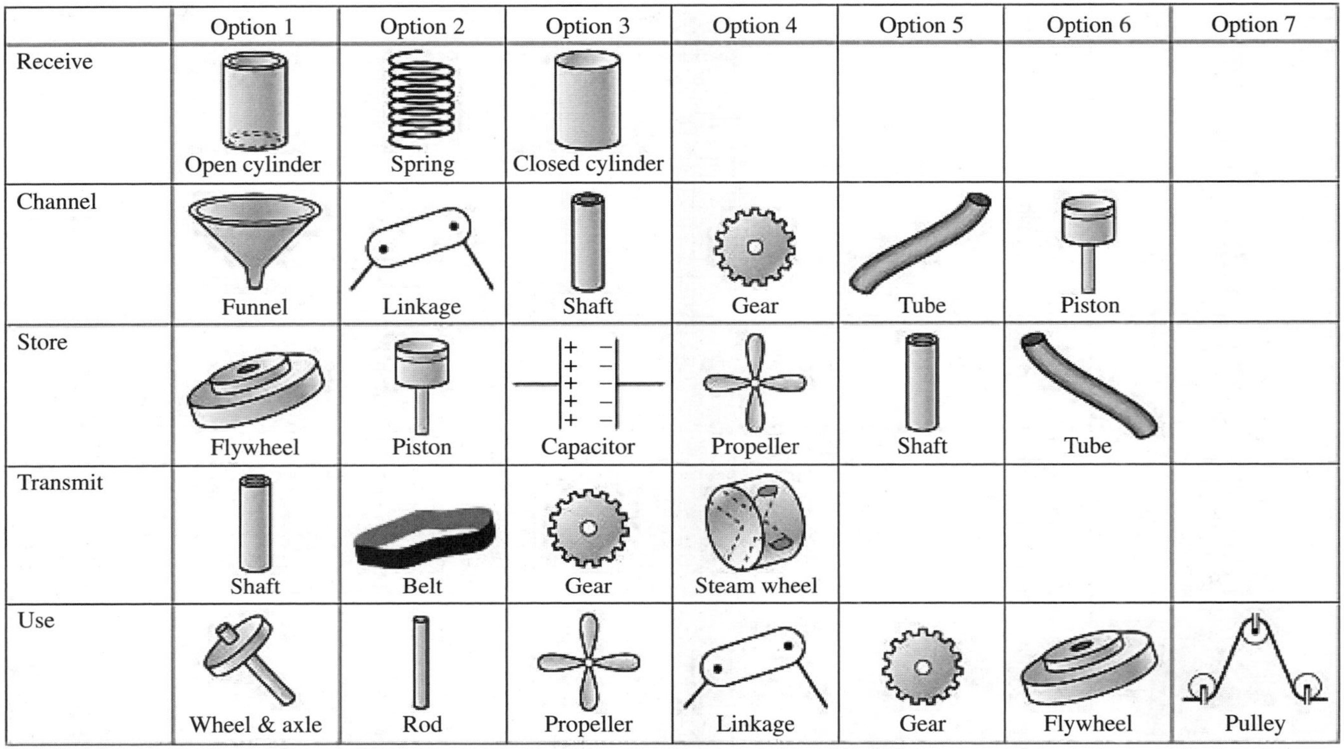

	Option 1	Option 2	Option 3	Option 4	Option 5	Option 6	Option 7
Receive	Open cylinder	Spring	Closed cylinder				
Channel	Funnel	Linkage	Shaft	Gear	Tube	Piston	
Store	Flywheel	Piston	Capacitor	Propeller	Shaft	Tube	
Transmit	Shaft	Belt	Gear	Steam wheel	Gear		
Use	Wheel & axle	Rod	Propeller	Linkage	Gear	Flywheel	Pulley

Figure 8.6 Morphological chart for machine shop kit.

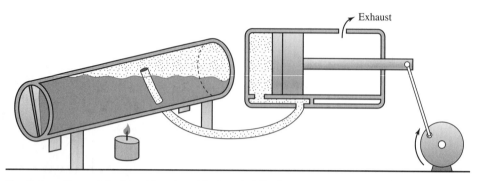

Figure 8.7   Concept I of machine shop kit: A tank full of water is heated to produce steam. The steam will travel through the tube and push the piston, which will turn the attached flywheel.

Figure 8.8   Concept II of machine shop kit: A flame is used to heat a piston, which will be pushed out to turn a gear. At the same time that the piston is pushed out, another piston is being pushed up, which will push the hot air piston back to its original position.

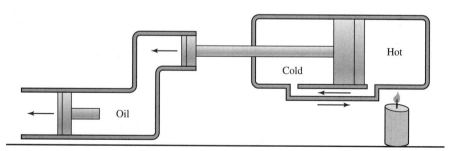

Figure 8.9   Concept III of machine shop kit: This design incorporates a system of pistons. The first piston is pushed by the pressure from heated air. It, in turn, compresses a medium of oil, which causes the final piston to be pushed.

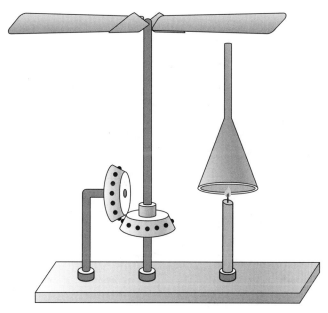

Figure 8.10 Concept IV of machine shop kit: Hot air is funneled to turn a propeller system. The propeller is connected to a central rod, which has a gear attached to it. The rotation of blades will cause the attached gear to rotate, which turns the other gear.

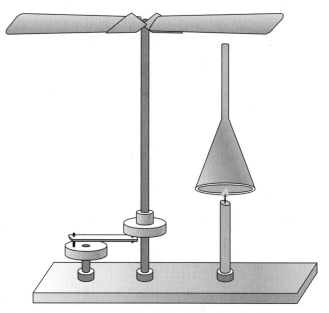

Figure 8.11 Concept V of machine shop kit: The hot air is channeled, which causes the propeller to rotate, which spins a flywheel. The flywheel is connected to a second flywheel by a connector link. Therefore, as the first flywheel turns, the second flywheel will also turn.

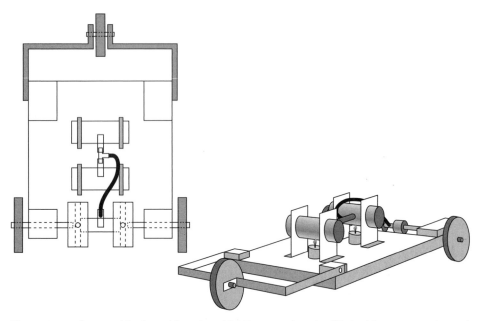

Figure 8.12   Concept VI of machine shop kit: Two metal tanks filled with water are heated with an alcohol burner. The heated water then generates steam that travels through a nylon tube to a steam tube. The steam tube is connected to two "steam wheels," which have holes drilled in them at 90° angles. The escaping steam will create rotation, which will turn the axles that turn the wheels and move the car.

Using the evaluation chart, each design was rated against a datum, the Stirling engine, which is currently used in the machine shop. The rating criteria were based on the design specification. A positive sign indicates that a design has a better rating than the datum. A negative sign, on the other hand, denotes that the design is less desirable than the datum in that particular category. If it so happens that the design is neither better nor worse at meeting a specification than the datum, then the value of zero is assigned in the chart. After carefully evaluating each design, the total number of positives and negatives was calculated. Furthermore, the weighted total was determined by multiplying each positive or negative by the specification weight and then adding them. Upon completing the design evaluations, it was concluded that concept VI best met the requirements and therefore was chosen as our project design. Figure 8.13 shows the evaluation table.

8.5   CONCEPTS EVALUATION: AUTOMATIC CAN CRUSHER

From the morphological chart, many possible combinations of devices can be obtained, but not all of these processes are positive or able to be accomplished effectively.

In this stage, the design team begins to limit the possible designs based on specifications, manufacturability, cost, and other factors. The surviving designs are then evaluated against each other. The concepts for the automatic can crusher are shown in

Evaluation Chart							
	Objective weight/10	Sketch 1	Sketch 2	Sketch 3	Sketch 4	Sketch 5	Sketch 6	D
Easy to assemble	7	0	0	0	+	0	+	A
Easy to disassemble	7	0	0	0	+	+	+	T
Safe for operator	10	0	0	0	0	0	0	U
Low vibration	5	+	−	+		0	0	M
Portable	4	−		0	0	0	+	
No sharp edges	6	+		0/+	−	−	0	
Retails for less than competition	9	+	+	+	+	+	+	
Convert energy efficiently	10	−	0	0	0	0	0	
No flying debris	8	0	0	0	0	0	0	
Low pollution	3	0	0	0	0	0	0	
Low replacement part cost	7	+	0	0	+	+	+	
Low noise	4	0	+	+	0	0	+	
Strong material	6	0	0	0	0	0	−	
Low energy dissipation	8	+	0	0	−	0	−	
Aesthetically appealing	5	−	0	−	0	0	+	
Total +		5	2	4	4	5	7	
Total −		3	1	1	2	1	2	
Overall total		2	1	3	2	4	5	
Weighted total		16	8	19	16	22	29	

Figure 8.13 Evaluation table for the machine shop kit.

Figures 8.14 through 8.20. A decision matrix is constructed to evaluate the designs based on the wishes obtained from the specification table. Specifications are weighted by importance, with 1 being the lowest and 10 being the highest. The concepts are evaluated, and one design is chosen by the design team to be the best possible design. This design is considered the datum by which all of the other designs are compared. Then, all of the other designs are evaluated against the datum and receive a positive sign [+] if they exceed the datum in the wish, a negative sign [−] if they are below the datum in the wish, and zero [0] if they are equal to the datum in the wish. The results are tallied, and scores are evaluated. Then the best design is placed as the datum, and the evaluation is done again. The highest scoring design is the chosen design to be built.

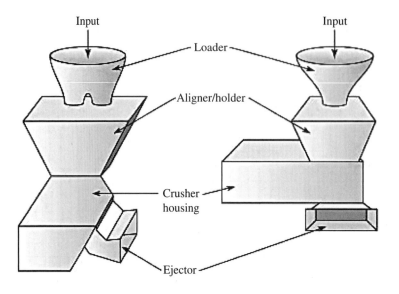

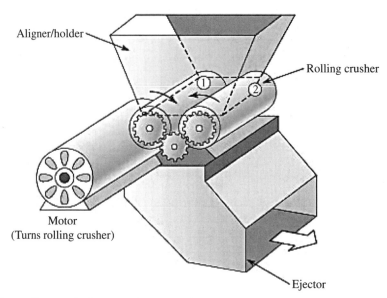

Figure 8.14 Concept I of automatic can crusher.

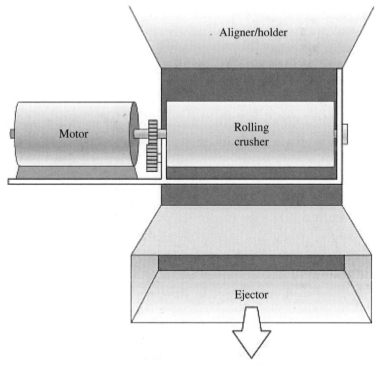

Figure 8.15   Concept II of automatic can crusher.

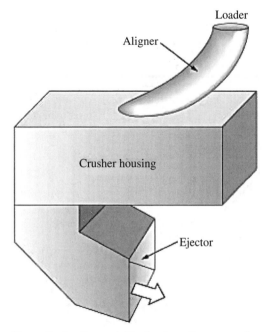

Figure 8.16   Concept III of automatic can crusher.

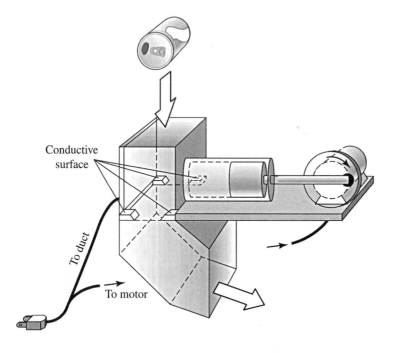

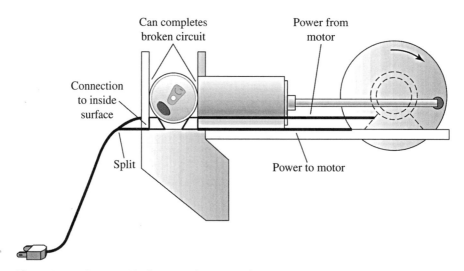

Figure 8.17 Concept IV of automatic can crusher.

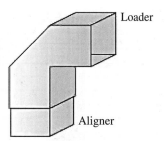

Loader

Aligner

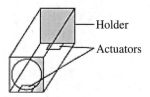

Holder

Actuators

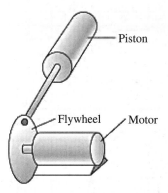

Piston

Flywheel

Motor

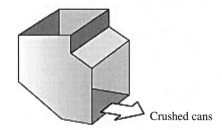

Crushed cans

Figure 8.18 Concept V of automatic can crusher.

Criterion	Wt	Concepts				
		I	II	III	IV	V
Inoperable when opened	10	0	0	0	0	
Runs on standard 110-V outlet	7	0	0	0	0	
<$50 retail	9	0	0	0	0	
Enough force to crush glass and plastic	3	0	0	0	0	D
Large storage ofcrushed cans	5	0	0	0	0	A
Easily accessible kill switch	3	0	0	0	0	T
No sharp corners	10	0	0	0	0	U
Easy to start	8	0	0	0	0	M
Stop easily and immediately	9	0	0	0	0	
Can stop in mid operation	8	0	0	0	0	
Drains residual fluid from machine	10	0	0	0	0	
Low vibration	8	0	0	0	0	
Low peripheral force	8	0	0	0	0	
Shock absorption	8	0	0	0	0	
High-efficiency engine	8	0	0	0	0	
High material strength	9	0	0	0	0	
Small force to depress switches	9	0	0	0	0	
Safety stickers	9	0	0	0	0	
Reset button	9	0	0	0	0	
No flying debris	9	0	0	0	0	
Operating steps strickers	9	0	0	0	0	
Low loading height	7	0	0	0	0	
Easily accessible interior	7	0	0	0	0	
Crushing mechanism inaccessible	10	−	0	−	0	
Crush many cans a minute	9	+	+	0	0	
Utilize gravity in design	4	0	0	0	0	
Low noise output	5	0	0	0	0	
Starts up immediately	5	0	0	0	0	
Easy access to clear jams	6	−	−	0	0	
Large capacity loader	5	+	+	0	0	
Long running capability	7	0	0	0	0	
Parts easy to acquire	8	−	−	0	0	
Internal part safe from liquid damage	9	0	0	0	0	
Low operating cost	5	0	0	0	0	
Easy cleaning	6	−	−	0	0	
Easy to disassemble	6	0	0	0	0	
Crushes along major axis	2	0	0	0	0	
No exhaust	2	0	0	0	0	
Total +		2	2	0	1	
Total −		4	3	1	0	
Overall total		−2	−1	−1	−1	
Weighted total		−16	−6	−10	−2	

Figure 8.19   Evaluation of designs: column V is datum.

		Concepts				
Criterion	Wt	I	II	III	IV	V
Inoperable when opened	10	0	0	0	0	
Runs on standard 110-V outlet	7	0	0	0	0	
<$50 retail	9	0	0	0	0	
Enough force to crush glass and plastic	3	0	0	0	0	D
Large storage of crushed cans	5	0	0	0	0	A
Easily accessible kill switch	3	0	0	0	0	T
No sharp corners	10	0	0	0	0	U
Easy to start	8	0	0	0	0	M
Stop easily and immediately	9	0	0	0	0	
Can stop in mid operation	8	0	0	0	0	
Drains residual fluid from machine	10	0	0	0	0	
Low vibration	8	0	0	0	0	
Low peripheral force	8	0	0	0	0	
Shock absorption	8	0	0	0	0	
High-efficiency engine	8	0	0	0	0	
High material strength	9	0	0	0	0	
Small force to depress switches	9	0	0	0	0	
Safety stickers	9	0	0	0	0	
Reset button	9	0	0	0	0	
No flying debris	9	0	0	0	0	
Operating steps strickers	9	0	0	0	0	
Low loading height	7	0	0	0	0	
Easly accessible interior	7	0	0	0	0	
Crushing mechanism inaccessible	10	−	0	−	0	
Crush many cans a minute	9	+	+	0	0	
Utilize gravity in design	4	0	0	0	0	
Low noise output	5	0	0	0	0	
Starts up immediately	5	0	0	0	0	
Easy access to clear jams	6	−	−	0	0	
Large capacity loader	5	+	+	0	0	
Long running capability	7	0	0	0	0	
Parts easy to acquire	8	−	−	0	0	
Internal part safe from liquid damage	9	0	0	0	0	
Low operating cost	5	0	0	0	0	
Easy cleaning	6	−	−	0	0	
Easy to disassemble	6	0	0	0	0	
Crushes along major axis	2	0	0	0	0	
No exhaust	2	0	0	0	0	
Total +		2	2	0	1	
Total −		4	3	1	0	
Overall total		−2	−1	−1	1	
Weighted total		−16	−6	−10	2	

Figure 8.20   Evaluation of designs: column IV is datum.

8.6 PROBLEMS

8.6.1   Team Activities

1. In order to enhance your engineering sketching ability, consider the paper stapler. Disassemble the stapler to its basic components. Sketch these components and then sketch an assembled stapler.
2. Why would a designer place items (demos of function) in the morphological chart if he or she knows that they would not be feasible?
3. Define engineering sense. If someone asks you to demonstrate a speed of 10 cm/sec, how would you do so using your engineering sense?
4. Why would you need to evaluate alternatives if your gut feeling is pointing you toward one of these alternatives?

8.6.2   Individual Activities

1. Discuss the following: "Using different alternatives for each function allows design engineers to substitute that particular element (in case of a problem) rather than changing the whole design."
2. What difference does it make if you use the absolute or relative scale in the weighing function? Which one would you recommend and why?
3. Why would you evaluate the different alternatives individually rather than as a team?
4. In all of examples presented in this chapter, the design team has the opportunity to establish the evaluation criteria (which thus may be biased by the design team). What would you use as mechanism to establish the evaluation criteria if you are the design manager at Ford Motor Company?
   a. How would you conduct the evaluation?
   b. How did the house of quality contribute to the evaluation mechanism?
5. In the event that the mechanism that satisfies a function, or its presentation, is found to have difficulty in attaching to another function(s), what would you do and why?
6. Name situations in which you have to perform the evaluation chart more than once.
7. Use the Pugh method to evaluate this course. Use your physics course as a datum.

8.7   Selected Bibliography

AMBROSE, S. A. and AMON, C. H. "Systematic Design of a First-Year Mechanical Engineering Course at Carnegie Mellon University." *Journal of Engineering Education*, pp. 173–181, 1997.

BURGHARDT, M.D. *Introduction to Engineering Design and Problem Solving*. New York: McGraw-Hill, 1999.

CROSS, N., CHRISTIAN, H., and DORST, K. *Analysing Design Activity*. New York: Wiley, 1996.

EEKELS, J. and ROOZNBURG, N. F. M. "A Methodological Comparison of Structures of Scientific Research and Engineering Design. Their Similarities and Differences." *Design Studies*, Vol. 12, No. 4, pp. 197–203, 1991.

PAHL, G. and BEITZ, W. *Engineering Design: A Systematic Approach*. New York: Springer-Verlag, 1996.

PUGH, S. *Total Design*. Reading, MA: Addison-Wesley, 1990.

SUH, N. P. *The Principles of Design*. New York: Oxford University Press, 1990.

ULMAN, D. G. *The Mechanical Design Process*. New York: McGraw-Hill, 1992.

ULRICH, K. T. and Eppinger, S.D. *Product Design and Development*. New York: McGraw-Hill, 1995.

VIDOSIC, J. P. *Elements of Engineering Design*. New York: The Ronald Press Co., 1969.

WALTON, J. *Engineering Design: From Art to Practice*. New York: West Publishing Company, 1991.

Embodiment Design

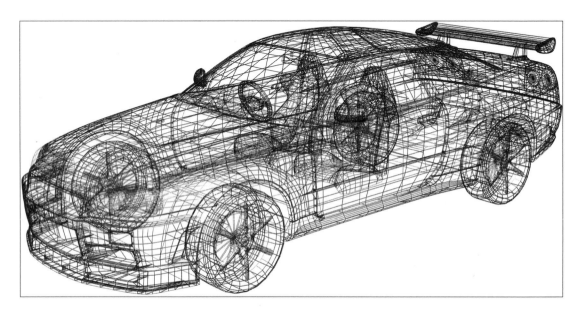

At this stage of the design process, the design team starts the embodiment design stage by present-ing information through concept sketches. These sketches are then fine tuned into product drawings. (Arogant/Shutterstock)

9.1 OBJECTIVES

By the end of this chapter, you should be able to

1. Discuss the different types of presentations of a product.
2. Discuss the difference between prototype and mock-up.
3. Construct a bill of material.
4. Understand the term design for "X"

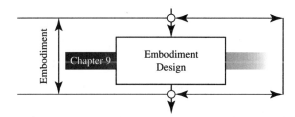

n the previous two chapters, different devices that perform the requirements were derived by using the morphological chart. The concepts were sketched, not in great detail, but in a way that presents the major functions that compose the device.

These concepts were then evaluated based on three different stages. The first stage involves subjecting the concept to the specification table. If it passes the demands, then a feasibility test is conducted. If the concept passes the first two stages, then it will be subjected to relative evaluation based on either the Pugh or decision matrix method. At this stage of the design process, the design team starts what can be called the embodiment design stage. The word *embody* means "to give something a tangible expression."

This chapter discusses presentation of the concept(s) that pass the evaluation criteria. The idea of presenting information through drawings was introduced when different concepts were generated from the morphological chart. The sketches that present the concepts are not detailed: They simply provide enough information to demonstrate the combination of the different function mechanisms that produce the concept.

It should be reemphasized that the design process is an iterative process. This means that, if additional refinements are found necessary (after the skeleton sketches or while in the evaluation stage), the designer should iterate back to the concept phase and possibly generate new concepts. The drawback of this process is that it takes time, but that time should be initially implanted in the scheduling of events. The use of functional structure and morphological charts makes it easy on the designer to replace components in the design rather than start over if a problem arises. The objective tree and the house of quality provide the evaluation criteria. These steps have well-defined boundaries and thus facilitate the iterative process.

In the world of engineering, information needs to be passed on through drawings. The analysis, manufacturing, and assembly of the product are based on drawings. The following section discusses the type of drawings that are required by the design team.

9.2 PRODUCT DRAWINGS

In the design process, drawings may be used to serve the following objectives.

1. Drawings are the preferred form of data communications in the world of engineering.
2. Drawings are used to communicate ideas between the designers themselves and between the designers and manufacturing personnel.
3. Drawings simulate the operation of the product.
4. Drawings are used to check for the completeness of the product.

There are different types of drawings that can be used at different levels.

1. *Sketches (layout drawing)—CONCEPTUAL DESIGN:* Sketches are generated during the conceptualization stage of the design process. Different mechanisms of each function or subfunction are first sketched. These sketches are then used to build devices. These sketches present the completeness of the device. They are working documents that support the development of the major components and their relations. They are made to scale, but only the important dimensions are shown in the layout. Tolerances are usually not shown unless they are critical. This form of drawing is needed in the analysis. When an individual component is analyzed, a sketch presents the relation of this part to its surroundings.
2. *Assembly drawings—EMBODIMENT DESIGN:* The goal of the assembly drawing is to show how components fit together. Figure 9.1 shows a typical assembly drawing. An assembly drawing has the following characteristics:
   a. Each component is identified with a number referenced to a list of all components.
   b. Necessary, detailed views are included to explain information that is not clear in other views.
3. *Detail drawings—DETAILED DESIGN:* As the product evolves, a detail of individual components develops. These types of drawings are carried out when all the dimensions have been calculated in the detailed design phase and will be discussed in the next chapter. A typical detail drawing is shown in Figure 9.2. The important characteristics of a detail drawing are
   a. all dimensions must include tolerances, and
   b. materials and manufacturing detail must be clear and presented in specific language.
   These drawings are made using a computer-aided design (CAD) system, such as Pro/E.

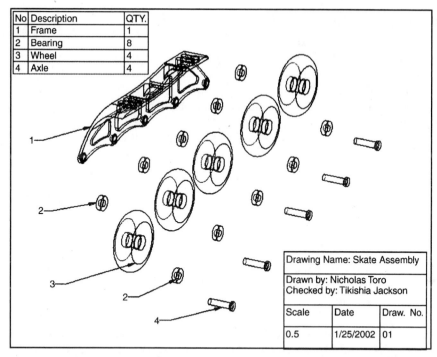

No	Description	QTY.
1	Frame	1
2	Bearing	8
3	Wheel	4
4	Axle	4

Drawing Name: Skate Assembly		
Drawn by: Nicholas Toro		
Checked by: Tikishia Jackson		
Scale	Date	Draw. No.
0.5	1/25/2002	01

Figure 9.1   Typical assembly drawing.

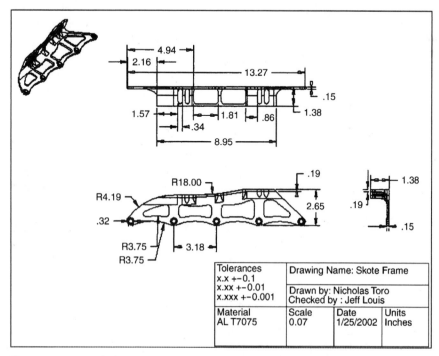

Tolerances	Drawing Name: Skote Frame		
x.x +−0.1			
x.xx +−0.01	Drawn by: Nicholas Toro		
x.xxx +−0.001	Checked by : Jeff Louis		
Material	Scale	Date	Units
AL T7075	0.07	1/25/2002	Inches

Figure 9.2   Typical detail drawing.

9.3 PROTOTYPE

The experimental phase of the engineering design requires the transformation of paper drawings into hardware, which is constructed and tested to verify the concept's workability.

Four construction techniques are available for designers.

1. *Mock-up:* This is generally constructed to scale from plastics, wood, cardboard, and so forth to give the designer a feel for his or her design. It is often used to check the clearance, assembly techniques, manufacturing considerations, and appearance. This is the least expensive technique, is relatively easy to produce, and can be used as a tool in selling the idea to clients or management. A solid model in the CAD system can often replace the mock-up. Mock-ups are usually referred to as a *proof of concept prototype.*

2. *Model:* The model relates the physical behavior of the system through mathematical similitude. The modeling is usually referred to as a proof of product prototype. There are different types of models used to predict the behavior of a real system.
   a. *True* is an exact geometric reproduction of the real system, built to scale, and satisfying all restrictions imposed by the design parameters.
   b. *Adequate* is constructed to test specific characteristics of the design and is not intended to yield information concerning the total design.
   c. *Distorted* purposely violates one or more design conditions. This violation is often required when it is difficult or impossible to satisfy the specific conditions due to time, material, or physical characteristics, and it is felt that reliable information can be obtained through the distortion.

3. *Prototype:* This is the most expensive technique and the one that produces the greatest amount of useful information. The prototype is a constructed, full-scale, and working physical system. Prototypes can be comprehensive or focused. A comprehensive prototype corresponds to a full-scale, fully operational version of the whole product. An example of a comprehensive prototype is the beta prototype, which is given to customers in order to identify any remaining design flaws before committing to production. In contrast, focused prototypes correspond to one or a few of the product elements. Examples of focused prototypes include foam models used to explore different forms of a product.

4. *Virtual prototyping:* Computer-aided design (CAD) and computer-aided engineering (CAE) software suppliers have fulfilled the goals of virtual prototyping through the delivery of 3-D feature-based modeling capabilities. Solid modeling has enabled the use of geometry to visualize product models quickly and to detect gross interference problems. One of the unique features in feature-based modeling software is its ability to automate design changes. It provides integrated capabilities for creating detailed solid and sheet-metal components, building assemblies, designing weldments, and producing fully documented production drawings and photorealistic renderings.

Feature-based modeling is built on combining the commands needed to produce a common feature. For example, a combination of commands to create a hole of prescribed dimensions and relations to keep the parts of the hole maintained is stored in a feature that is given the name *hole*.

Parametric modeling allows the presentation of dimensions of the parts in term of parameters. Virtual prototyping can have its biggest payoff in manufacturing and analysis. Using solid models as the base for a finite-element analysis tool would reduce prototyping time and save money. In manufacturing, parametric feature-based models will reduce the time needed to produce a feasible prototype.

9.4 DESIGN FOR "X"

Unlike smaller engineering design projects, in an industrial setting, many teams contribute to create new products including non-engineering teams. Several attributes are put into the product to keep up with competition and to produce reliable, marketable, safer, and less costly products in shorter periods of times. In order to integrate these attributes through various activities and maintain cost-effective benefits, a process called *concurrent engineering* has been developed. The use of this approach implies systems theory thinking. Synergistic results become the main goal of its processes.

Designing for such different attributes (such as manufacturability, assembly, environment, safety, etc.) are usually referred to as a design for "X".

9.4.1 Design for Manufacturing

Design for manufacturing (DFM) is based on minimizing the costs of production, including minimizing the time to market while maintaining a high standard of quality for the product. DFM provides guidance in the selection of materials and processes and generates piece-part and tooling-cost estimates at any stage of product design. You can find several companies on the Internet that sell products and software to facilitate the functioning of a company to perform elements for DFM.

This would include:

1. an accurate cost estimator that reviews the cost of parts as they are being designed in a fast and accurate way;
2. a concurrent engineering implementation that provides quantitative cost information which allows the design team to make decisions based on realtime information and to shorten the product development time;
3. provide supplier negotiations with unbiased details of cost drivers; and
4. competitive benchmarking that compares the designs with competitors' products to determine marketability and target cost.

There are several methods that have been used in DFM to assess in analysis, such as process-driven design, group technology, failure mode and effect analysis, value engineering, and the Taguchi method.

9.4.2 Design for Assembly

Design for assembly (DFA) is the study of the ease of assembling various parts and components into a final product. A lower number of parts and an ease of assembly contribute to reducing the overall cost of the product. With DFA, every part has to be checked. It must be determined if it is a necessary part, if it would be better integrated into other parts, or if it should be replaced by a similar function part that is simpler and costs less. Integrating both design for manufacturing and design for assembly helps contribute to the competitive success of any given product by matching that product's demands to its manufacturability and assembly capabilities.

9.4.3 Design for Environment

Recently, a trend of environmentally conscious products along with the rapid implementation of worldwide environmental legislation have put the responsibility for the end-of-life disposal of products on to the manufacturer. Manufacturers must now implement rules during the design that enforce *design for environment* (DFE). By designing products up-front for environmental and cost efficiency, manufacturers are gaining an edge on slow-to-react competitors who face the same issues unprepared. DFE is concerned with the disassembly of products at the end of their useful life and reveals the associated cost benefits and environmental impacts of a product's design. This quantitative information then can be used to make informed decisions at the earliest concept stages of the product's design.

9.5 SAFETY CONSIDERATIONS

One very important consideration in engineering design maintains that the resulting product is safe for humans. In recent years, the liability decisions made by the courts further increased the importance of safety. According to the National Safety Council, over 105,000 Americans were accidentally killed and over 10 million suffered a disabling injury in 1980, which translated into $83.2 billion in disability payments. Safety is hardly new; it has been important for thousands of years. There are many safety functions used during the design process, including

1. developing accident prevention requirements for the basic design;
2. participating in the design reviews; and
3. performing hazard analyses during the product design cycle.

9.5.1 Safety Analysis Techniques

The two techniques used for design stage and reliability analysis—failure modes and effect analysis (FMEA) and fault trees—can also be applied for safety analyses.

Failure Modes and Effect Analysis

This method was originally developed for use in the design and development of flight control systems. The method can also be used to evaluate design at the initial stage from the point of view of safety. Basically, the technique calls for listing the potential failure modes of each part, as well as the effects on the parts and on humans. The technique may be broken down into seven steps.

1. Defining system boundaries and requirements.
2. Listing all items.
3. Identifying each component and its associated failure modes.
4. Assigning an occurrence probability or failure rate to each failure mode.
5. Listing the effects of each failure mode on concerned items and people.
6. Entering remarks for each possible failure mode.
7. Reviewing and initiating appropriate corrective measures.

Fault Trees

This technique uses various symbols. It starts by identifying an undesirable event, called the top event, and then successively asking, "How could this event occur?" This process continues until the fault events do not require further development. If occurrence data are known for the basic or primary fault event, the occurrence measure for the top event can be calculated.

9.6 HUMAN FACTORS

The science that considers humans and their reactions to familiar, as well as strange, environments is known by several terms, including *human engineering*, human performance, and *ergonomics*. Adapting a human to fit a given environment is almost impossible, so methods for adapting the environment to suit the human must be devised. Human behavior is critical to the success of an engineering system. Therefore, it is important to consider typical human behaviors during the design phase. Examples of typical human behaviors include the following.

1. People are often reluctant to admit errors.
2. People usually perform tasks while thinking about other things.
3. People frequently misread or overlook instructions and labels.
4. People often respond irrationally in emergency situations.
5. A significant percentage of people become complacent after successfully handling dangerous items over a long period of time.

6. Most people fail to recheck outlined procedures for errors.

7. People are generally poor estimators of speed, clearance, or distance. They frequently overestimate short distance and underestimate large distance.

8. People are generally too impatient to take the time needed to observe.

9. People are often reluctant to admit that they cannot see objects well, due either to poor eyesight or to inadequate illumination. They generally use their hands for examining or testing.

10. Every time people dial telephones, drive cars, study oscilloscopes, use computers, or form parts on a lathe, they have joined their sensing, decision-making, and muscular powers into an engineering system. Figure 9.3 shows the coupling of components that are necessary to consider a human as an integral part of an efficient human-machine system.

9.6.1 Human Sensory Capabilities

Human sensory capabilities include the following.

1. *Sight:* Human eyes see differently from different angles or positions. For example, looking straight ahead, the eyes can perceive all colors. However, with an increase in the viewing angle, human perception begins to decrease. The limits of color vision are given in Table 9.1. Also, at night or in the dark,

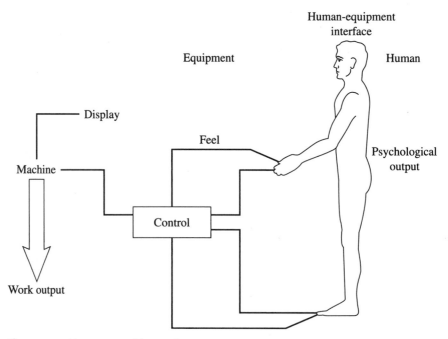

Figure 9.3 Human—machine system.

TABLE 9.1 Human Color Vision

Situation	Green	Blue	Yellow	White	Green-Red	Red
Vertical	40°	80°	95°	130°	—	45°
Horizontal	—	100°	120°	180°	60°	—

small-sized orange, blue, green, and yellow lights are impossible to distinguish from a distance.

2. *Noise:* The performance quality of a task requiring intense concentration can be affected by noise. It is an established fact that noise contributes to people's feelings, such as irritability, boredom, and well-being. A noise level of 90 dB is considered harmless, while above 100 dB is unsafe. Levels in excess of 130 dB are considered unpleasant and may actually become harmful.

3. *Touch:* Touch adds to, or may even replace, the information transmitted to the brain by eyes and ears. For example, it is possible to distinguish knob shapes with touch alone. This ability could be valuable if, for instance, the power goes out and there is no light.

4. *Vibration and motion:* It is an accepted fact that poor performance of physical and mental tasks by equipment operators could be partially or fully attributed to vibrations. For example, eye strain, headaches, and motion sickness could result from low-frequency, large-amplitude vibrations.

5. Useful guidelines for the effects of vibration and motion are as follows.
   a. Eliminate vibrations with an amplitude greater than 0.08 mm.
   b. Use such devices as shock absorbers and springs wherever possible.
   c. Use damping materials or cushioned seats to reduce vibration wherever possible.
   d. Seated humans are most affected by vertical vibration, so use this information to reduce vertical vibrations.
   e. The resonant frequency of the human vertical trunk, in the seated position, is between three and four cycles per second. Therefore, avoid any seating that would result in or would transmit vibrations of three to four cycles per second.

9.6.2 Anthropometric Data

Anthropometry is the science that deals with the dimensions of the human body.

Anthropometric data are divided into statistical groups known as percentiles. If 100 people are lined up from smallest to largest in any given respect, they would be classified from the 1 percentile to the 100 percentile. The 2.5 percentile means that designs based on this series of dimensions would allow up to 2.5% of the population to use the system and the remaining 97.5% would be excluded. Figure 9.4 shows the dimensions of a standing adult male.

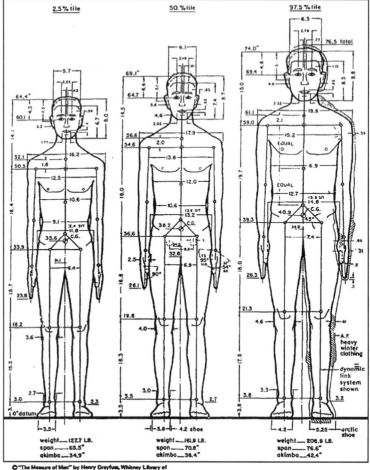

Figure 9.4 Standing adult male. (Reprinted by permission of Henry Dreyfuss Associates.)

LAB 9: Ergonomics

Purpose

This lab provides an introduction to some of the ergonomics issues in design.

Worked Example

Table LT9.1 shows a series of established standards.[1] The following Web page shows detailed drawings of the dimensions: http://www.openerg.com. Refer to Table LT9.1 for this example.[2]

---

[1]Data from http://cad.fk.um.edu.my/ergonomics/Anthropometry/US.pdf
[2]This example is adapted from Niebel, B. and Freivalds, A. *Methods Standards and Work Design*. New York: McGraw-Hill, 1999.

TABLE LT9.1 Anthropometric measured data in mm of U.S. adults, 19 to 60 years of age. According to Gordon, Churchill, Clauser, et al. (1989), who used the numbers in brackets

Dimension	Men 5th %ile	Mean	95th %ile	SD	Women 5th %ile	Mean	95th %ile	SD
1. Stature	1647	1756	1867	67	1528	1629	1737	64
2. Eye height, standing [D19]	1528	1634	1743	66	1415	1516	1621	63
3. Shoulder height (acromion), standing [2]	1342	1443	1546	62	1241	1334	1432	58
4. Elbow height, standing [D16]	995	1073	1153	48	926	998	1074	45
5. Hip height (trochanter) [107]	853	928	1009	48	789	862	938	45
6. Knuckle height, standing	Na	Na	Na	Na	Na	Na	Na	Na
7. Finger height, standing [D13]	591	653	716	40	551	610	670	36
8. Sitting height [93]	855	914	972	36	795	852	910	35
9. Sitting eye height [49]	735	792	848	34	685	739	794	33
10. Sitting shoulder height (acromion) [3]	549	598	646	30	509	556	604	29
11. Sitting elbow height [48]	184	231	274	27	176	221	264	27
12. Sitting thigh height (clearance) [104]	149	168	190	13	140	160	180	12
13. Sitting knee height [73]	514	559	606	28	474	515	560	26
14. Sitting popliteal height [86]	395	434	476	25	351	389	429	24
15. Shoulder-elbow length [91]	340	369	399	18	308	336	365	17
16. Elbow-fingertip length [54]	448	484	524	23	406	443	483	23
17. Overhead grip reach, sitting [D45]	1221	1310	1401	55	1127	1212	1296	51
18. Overhead grip reach, standing [D42]	1958	2107	2260	92	1808	1947	2094	87
19. Forward grip reach [D21]	693	751	813	37	632	686	744	34
20. Arm length, vertical [D3]	729	790	856	39	662	724	788	38
21. Downward grip reach [D43]	612	666	722	33	557	700	664	33
22. Chest depth [36]	210	243	280	22	209	239	279	21
23. Abdominal depth, sitting [1]	199	236	291	28	185	219	271	26
24. Buttock-knee depth, sitting [26]	569	616	667	30	542	589	640	30
25. Buttock-popliteal depth, sitting [27]	458	500	546	27	440	482	528	27
26. Shoulder breadth (biacromial) [10]	367	397	426	18	333	363	391	17
27. Shoulder breadth (bideltoid) [12]	450	492	535	26	397	433	472	23
28. Hip breadth, sitting [66]	329	367	412	25	343	385	432	27
29. Span [98]	1693	1823	1960	82	1542	1672	1809	81
30. Elbow span	Na	Na	Na	Na	Na	Na	Na	Na
31. Head length [62]	185	197	209	7	176	187	198	6
32. Head breadth [60]	143	152	161	5	137	144	153	5
33. Hand length [59]	179	194	211	10	165	181	197	10
34. Hand breadth [57]	84	90	98	4	73	79	86	4
35. Foot length [51]	249	270	292	13	224	244	265	12
36. Foot breadth [50]	92	101	110	5	82	90	98	5
37. Weight (kg), estimated by Kroemer	58	78	99	13	39	62	85	14

Ergonomics is defined as "the study of the problems of people adjusting to their work environment." Engineers and designers must keep the effects on human users forefront in their product designs. *(Bonsai/Shutterstock)*

The problem at hand here is to arrange seating in an auditorium such that most individuals will have an unobstructed view of the speaker screen.

Step 1.
Determine the body dimensions that are critical in the design (design criteria). Refer to Figure L9.1.
   a. Erect sitting height
   b. Seated eye heights

Step 2.
Define the population being served: U.S. adult males and females.

Step 3.
Select a design principle and the percentage of population to be accommodated: Designing for extremes accommodating 95% of the population. The key principle is to allow a 5th percentile female sitting behind a 95th percentile male to have an unimpeded line of sight.

Step 4.
Find values from Table LT9.1: 5th percentile female seated eye height is 685 mm; 95th percentile male sitting erect is 972 mm. Difference (972 − 685 = 287 mm). For a small female to see over a large male, there must be a rise of 287 mm between two rows. This creates a large slope. Suggestions?

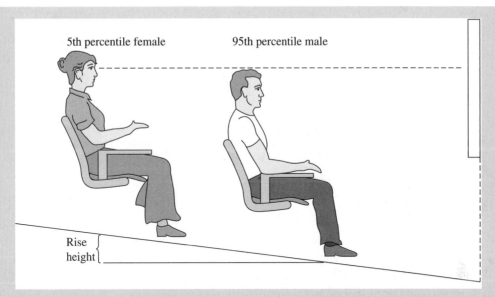

Figure L9.1   Design criteria. (METHODS, STANDARDS, AND WORK DESIGN by FREIVALDS, ANDRIS. Copyright 1998 by MCGRAW-HILL COMPANIES, INC.—BOOKS. Reproduced with permission of MCGRAW-HILL COMPANIES, INC.—BOOKS in the format Textbook via Copyright Clearance Center.)

Step 5.
Optimization allows for heavy clothes and shoes. Most measurements are made for a naked human.

Lab 9 Problems

1. Measure the dimensions of your seat in the classroom. Redesign the seating to fit 95% of females. How would the design affect 50% of males?

2. In a computer lab, suggest a design that will allow most of the users to view the PC and to view their classmates, who are sitting in different rows.

3. What would be the dimensions of a stage podium for your instructor?

4. Because of the *Challenger* disaster, NASA has decided to include a personal escape capability (i.e., a launch compartment) for each space shuttle astronaut. Because space is at a premium, proper anthropometric design is crucial. Also, because of budget restrictions the design is to be nonadjustable (for example, the same design must fit all present and future astronauts, both male and female). For each launch compartment feature (Figure L9.2), indicate the body feature used in the design, the design principle used, and the actual value (in inches) to be used in its construction.

   a. Height of seat
   b. Seat depth
   c. Height of joystick
   d. Height of compartment
   e. Depth of foot area
   f. Depth of leg area

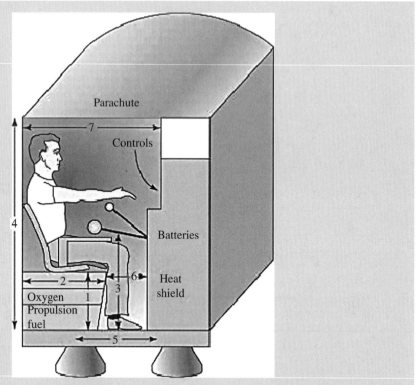

Figure L9.2   Compartment. (METHODS, STANDARDS, AND WORK DESIGN by FREIVALDS, ANDRIS. Copyright 1998 by MCGRAW-HILL COMPANIES, INC.—BOOKS. Reproduced with permission of MCGRAW-HILL COMPANIES, INC.—BOOKS in the format Textbook via Copyright Clearance Center.)

    g. Depth of chamber

    h. Width of compartment

    i. Weight limit

5. Choose the captain of your team to measure and record his or her anthropomorphic dimensions on the chart shown in Figure L9.3. In what percentile would you classify your captain?

6. Discuss four major classifications of materials.

7. List the classifications of manufacturing processes.

8. Describe the following:

    (a) Casting

    (b) Forging

    (c) Welding

9. Describe the following terms:

    (a) Ergonomics

    (b) Human error

    (c) Human reliability

    (d) Human engineering

    Hint: Search in the library for more information.

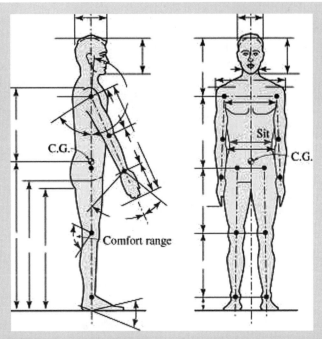

Figure L9.3 Anthropometric chart. (Reprinted by permission of Henry Dreyfuss Associates.)

10. Consider that you have been assigned the task of designing the environment for an industrial worker who must be seated at a control console for four continuous hours each day. The worker monitors and controls plant and laboratory temperature and humidity. Describe (list) the types of environment that would be considered ideal for this operator.

11. Design an optimum display of controls, indicators, and utility items listed in the following proposed dash layout (see also Figure L9.4), paying particular attention to human

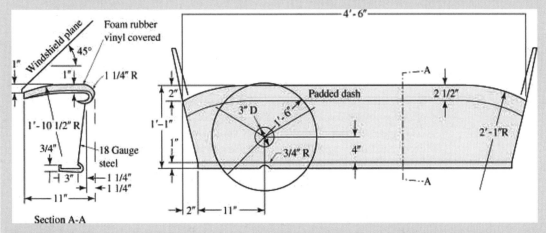

Figure L9.4 Proposed dash design layout. (From THE SCIENCE OF ENGINEERING, 1st edition, by P.H. Hill © 1970.)

engineering principles. Where present dials and indicators are ineffective, redesign the display of information for efficient use by the human operator. Ask yourself the following questions about the information to be conveyed: "Which is the most important?" "Least important?" "Which is the most often used?" "Least often used?" "What information must the indicator convey to the operator?" "Does it provide this information?" "Does it overinform?" "Underinform?" "What medium does it use?" Most indicators are used visually, and eyes are often overburdened in their functional usage. This would be critical in a high-speed car. Could the same information be obtained by other means, such as light, sound, or touch?

	Indicators	Utility Items
	Speedometer	Glove compartment
	Odometer	Radio
	Fuel	Cigarette lighter
	Engine temperature	Ventilation
Signals	Battery	Rear view mirror
	Oil pressure	
(Starter)	High-beam lights	

9.7 PROBLEMS

9.7.1 Team Activities

1. Consider a regular stapler and a hole puncher.
   a. Draw a function tree.
   b. Sketch the individual elements that were used to perform the functions.
   c. Sketch the overall device (combine the elements to obtain the full device).
   d. Construct a limited bill of materials for the device.

2. Survey your school CAD software and list the capabilities of each software. Some of this software has a very small learning curve, which may allow you to produce a 3-D model of your product in a very short time. If you have engineering graphics experience from a previous course, use that experience to check if the software has an analysis module that allows you to feed your model into the analysis module. Check the requirements of the analysis module.

3. In most architects' offices you will find a layout of future or current projects. What can you call these layouts?

9.8   Selected Bibliography

DEVON, R. and ENGLE, R. S. "The Effect of Solid Modeling Software on 3D Visualization Skill." *The Engineering Design Graphics Journal*, Spring, 1994.

EDWARDS, B. *Drawing on the Artist Within*. New York: Simon & Schuster Inc., 1986.

HAIK, Y. and KILANI, M. *Essentials of Pro/E*. Pacific Gove, CA: Brooks/Cole, 2001.

HARRISBERGER, L. *Engineersmanship: The Doing of Engineering Design*. Belmont, CA: Wadsworth, Inc., 1982.

JURICIC, D. and BARR, R. E. "The Place of Engineering Design Graphics in the Evolving Design Paradigm." *Proceedings of the ASEE Annual Conference*, Alberta, Canada, 1994.

NEE, J. G. "Freshman Engineering Design/Graphics Problem Status: A National Study." *The Engineering Design Graphics Journal*, Autumn, 1994.

RODRIGUEZ, W. *Visualization*. New York: McGraw-Hill, 1990.

ROORDA, J. "Visual Perception, Spatial Visualization and Engineering Drawing." *The Engineering Design Graphics Journal*, Spring, 1994.

SUH, N. P. *The Principles of Design*. Oxford University Press, 1990.

ULMAN, D. G. *The Mechanical Design Process*. New York: McGraw-Hill, 1992.

ULRICH, K. T. and EPPINGER, S. D. *Product Design and Development*. New York: McGraw-Hill, 1995.

Detailed Design

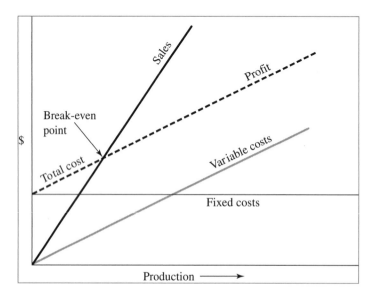

This break-even chart is designed to show, graphically, the profits and losses on the selling price and manufacturing costs of a product.

10.1 OBJECTIVES

By the end of this chapter, you should be able to

1. Understand the detailed design stage.
2. Identify and select engineering materials that suit a product.
3. Construct a bill of materials.
4. Use techniques introduced in this chapter to evaluate and analyze design cost.

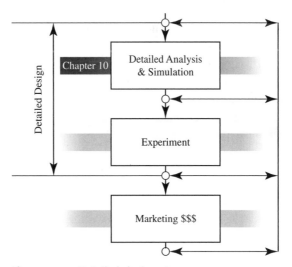

Figure 10.1 Detailed design stage.

The detailed design stage is the final step in the engineering design process before manufacturing and production is commenced. Most engineering degree courses will be within the detailed design stage framework. During this stage, commonly referred to as analysis and simulation, the designer selects the appropriate materials for each part and calculates accurately the dimensions and tolerances of the product. This process may include calculating variables such as static/dynamic loads, stresses, forces, temperature, pressures, fluid dynamics, electric current, resistance, chemical reactions, etc. where necessary. At this stage, the designer will also apply a suitable factor of safety to the design to ensure that the minimum requirements for non-failure of the product are well within the design limits. Obviously, each of these variables are subject areas in their own right and are beyond the scope of this book, however an introduction to the first phases of detailed design are summarized in this chapter.

10.2 ANALYSIS

Once the product is well defined, a complete analysis is conducted. In the analysis process, you need to evaluate the integrity of the design in terms of its safety concerns, choose the material that fits the requirements, and compute the cost involved.

Analysis proceeds as follows.

Step 1. Check for the design safety in terms of computing the forces that act on each component. Evaluate the stresses associated with these components.

Step 2. List the materials that would satisfy the stress requirements. Students are assumed to have some physical background from their physics classes before this course. It is assumed that the concept of force is known. The following section on material selection introduces the concept of stress.

Step 3. Students at this stage are also encouraged to seek experts' opinion on the analysis. Software that incorporates the CAD model into analysis mode is available. Students should investigate the availability of such software at their institution.

Steps 1, 2, and 3 are iterative until optimum results are achieved.

Step 4. Manufacturability of the components needs to be checked. Having a sketch of a component doesn't mean that the machinist will be able to produce the component. You need to provide the machinist with engineering drawings before the product component can be made. The engineering drawings should be detailed. Use CAD software to produce your drawings.

Step 5. Cost analysis of the components (obtained from a vendor or produced at the machine shop) needs to be conducted. You will be able to vary among components listed in the morphological chart if certain components are cost effective or fit the product better.

Step 6. The aesthetics or the "look" of the product needs to be maintained. Several customer surveys have shown that customers value the look of the product much more highly than its performance.

10.3 MATERIAL SELECTION

The design, even of the simplest element, requires the selection of a suitable material and a decision regarding the methods of manufacturing to be used in producing the element. These two factors are closely related, and the choice will affect the shape, appearance, cost, and so on. It may also determine the difference between a commercial success and a commercial failure. As the design becomes more complex and involves more elements, the selection of suitable materials and methods of production becomes more difficult. The design engineer must be sufficiently familiar with the characteristics and properties of the materials and the way in which they can be shaped to ensure that his or her decisions are well made.

10.3.1 Material Classifications and Properties

The various types of materials that may be used in a product design are classified as follows.

1. *Metals:* This important materials classification can be further divided into ferrous and nonferrous alloys. Ferrous alloys are based on iron. Nonferrous alloys are based on materials other than iron, such as copper, tin, aluminum, and lead.

2. *Ceramics and glass:* These are the result of the combination of metallic and nonmetallic elements. They are good insulators, brittle, thermally stable, and more wear resistant compared with metals. They are also harder and lower in thermal expansion than most metals.

3. *Woods and organics:* These are obtained from trees and plants. Their major advantage is that they are a renewable resource. However, some of their drawbacks include that they absorb water, require special treatment to prevent rotting, and are more flammable.

4. *Polymers or plastics:* These materials change viscosity with variation in temperature; therefore, they are easy to mold into a given shape. There are several benefits of polymers: They are good insulators; they are resistant to chemicals and water; they have a smooth surface finish; and they are available in many colors, which eliminates the need for painting. There are several drawbacks to polymers, however: They are low in strength; they deteriorate in ultraviolet light; and they have excessive creep at all temperatures.

The properties of materials can be divided into six categories: mechanical, thermal, physical, chemical, electrical, and fabrication. *Mechanical properties* include fatigue, strength, wear, hardness, and plasticity. *Thermal properties* include absorptivity, fire resistance, and expansion. *Physical properties* include permeability, viscosity, crystal structure, and porosity. *Chemical properties* include corrosion, oxidation, and hydraulic permeability. Four important *electrical properties* are hysteresis, conductivity, coercive force, and the dielectric constant. There are several *fabrication properties*, including weldability, castability, and heat treatability. A partial list of selected material properties is provided in Table 10.1

10.3.2 Material Selection Process

Several systematic approaches have been developed for selecting materials; one of these is given here.

Step 1. *Perform a material requirements analysis:*
This step is concerned with determining environmental and service conditions under which the product will have to operate.

Step 2. *List the suitable materials:*
This step calls for filtering through the available materials and obtaining several suitable candidates.

TABLE 10.1 Material Properties

Material	Density Kg/m³	ν Poisson's ratio	E MPa	σ_y MPa	σ_{ult} MPa	Melting Temperature °C	Thermal Conductivity W/mK
Pure Metals							
Beryllium	1827		3.033	3.792	6.205	1282	218
Copper	8858	0.33	1.172	0.689	2.206	1082	400
Lead	11349	0.43	0.138	0.138	0.172	327	35
Nickel	8858		2.069	1.379	4.826	1440	91
Tungsten	19 376		3.447		20.68	3367	170
Aluminum	2768	0.33	0.689	0.241	0.758	649	235
Alloys							
Aluminum 2024-T4	2768	0.33	0.731	3.034	4.137	579	121
Brass	8581	0.35	1.034	4.136	5.102	932	109
Cast iron (25T)	7197	0.2	0.896	1.655	8.274	1177	55
Steel: 0.2% C							
Hot Rolled	7833	0.27	2.068	2.758	4.826	1516	89
Cold Rolled	7833	0.27	2.068	4.482	5.516	1516	89
Stainless Steel Type 302 C. R.	7916	0.3	1.999	6.895	9.653	1413	21
Ceramics							
Crystalline Glass	2491	0.25	0.862	1.379		1249	1.3
Fused Silica Glass	2214	0.17	0.724			1582	1.1
Plastics							
Cellulose Acetate	1301	0.4	0.017	0.345	1.379		0.17
Nylon	1135	0.4	0.028	0.552	0.896		0.25
Epoxy	1107		0.045	0.483	2.068		0.35

Step 3. *Choose the most suitable material:*
This step involves analysis of the material listed, with respect to such factors as cost, performance, availability, and fabrication ability, and then selecting the most suitable material.

Step 4. *Obtain the necessary test data:*
This step calls for determining the important properties of the selected material, experimentally under real-life operational conditions.

Step 5. *Product specification fulfillment:*
The selected material must satisfy the stated specifications.

Step 6. *Cost:*
This important factor plays a dominant role in the marketing of the end product.

Step 7. *Material availability:*
The selected material must be available at a reasonable cost.

Step 8. *Material joining approach:*
Under real-life conditions, it may be impractical to produce an element using a single piece of material. This may call for manufacturing the components with different pieces of material joined together to form a single unit.

Step 9. *Fabrication:*
Engineering products usually require some level of fabrication, and different fabrication techniques are available. Factors affecting the fabrication method selected include time constraints, material type, product application, cost, and quantity to be manufactured.

Step 10. *Technical issues:*
Technical factors are mainly concerned with the material mechanical properties. Examples include strength compared with anticipated load, safety factors, temperature variation, and potential loading changes.

10.3.3 Primary Manufacturing Methods

The primary manufacturing methods, which are used to convert a material into the basic shape required, are as follows.

1. *Casting:* Casting is a widely used first step in the manufacturing process. During casting, an item takes on its initial usable shape. In the casting process, a solid is melted down and heated to a desirable temperature level. The molten material is then poured into a mold made in the required shape. The cast items may range in size and weight from a fraction of an inch to several yards. Typical examples include a zipper's individual teeth and the stern frames of ships.

2. *Forging:* Forging, which is among the most important methods of manufacturing items of high-performance uses, involves changing the shape of the piece of material by exerting force on that piece. The methods of applying pressure include the mechanical press, hydraulic press, and drop hammer. Products such as crankshafts, wrenches, and connecting rods are the result of forging. The material to be forged may be hot or cold.

3. *Machining:* Machining involves removing unwanted material from a block of material according to given specifications, such as size, shape, and finish, to produce a final product. There are many machining processes, such as milling, boring, grinding, and drilling.

4. *Welding:* Welding is a versatile production process that is used for combining items produced by some other means of manufacturing. Welding is the process of permanently joining two materials through coalescence, which involves a combination of pressure and surface conditions.

10.4 MATERIAL SELECTION THEORY—AN INTRODUCTION

10.4.1 Density

Density is one of the most important material properties because it determines the weight of the component. The range of material densities is based on the atomic mass and the volume the material occupies. Metals are dense because they consist of heavy and closely packed atoms. Polymers are formed from light atoms (carbons and hydrogen) and are loosely packed.

10.4.2 Melting Point

The melting point of a material is directly proportional to the bond energy of its atoms. Melting point becomes an important factor when the operating conditions of the material are in high-temperature environments, as in internal combustion engines and boilers. In design, the rule that generally is used is that the environment temperature should be 30% lower than the melting point of the material $((T/Tm < 0.3))$. Designers should also keep in mind that when a material is subjected to high temperature and undergoes stress, creep occurs (slow change of dimensions over time).

10.4.3 Coefficient of Linear Thermal Expansion

The coefficient of linear expansion is generally defined as

$$\alpha = \frac{dL}{LdT}$$

where
    L is the linear dimension of the object
    dL is the change in the linear dimension of the object
    dT is the change in temperature that causes the change in length

10.4.4 Thermal Conductivity

The thermal conductivity constant is a property of the material by which one can determine whether the material is a good heat conductor or insulator. In metals, electrons are the carriers of heat. A relation between the electrical conductivity and heat conductivity is established as

$$T = 5.5 \times 10^{-9} \frac{\text{colohm}}{\text{sec } K^2}$$

where
    k is the thermal conductivity
    s is the electrical conductivity
    T is the temperature in Kelvin.

10.4.5 Strength of Material

The strength of the material determines the amount of load a material can sustain before it breaks. In general, the design criterion is based on the yield stress of the material. The strength of the material is measured experimentally using a tensile testing machine. In a design-testing machine, a specimen is pulled with a certain load and the elongation that is associated with the load is measured. The load is converted to stress and the elongation is converted into strain. There are two general representations of the data: the stress–strain diagram and the true-stress–true strain diagram. In the stress–strain diagram, the stress is measured by

$$s = \frac{\text{load}}{\text{initial area}} = \frac{P}{A_o}$$

and the strain by

$$e = \frac{\text{change in length}}{\text{initial length}} = \frac{\Delta L}{L_o}$$

In the true-stress-true-strain diagram, the change in area after each loading is recognized and the true stress and true strains are obtained by

$$\text{stress} = \frac{\text{load}}{\text{instantaneous area}} = \frac{P}{A_{\text{inst}}}$$

$$E = \int_{L_D}^{L} \frac{dL}{L} = \ln\frac{L}{L_o}$$

The instantaneous area can be obtained by conservation of volume as

$$A_o L_o = AL = \text{constant}$$

Once the diagram is obtained, the modulus of elasticity can be obtained by finding the slope of the straight line before the yield stress.

Designers apply a safety factor based on the application and strength of the material.

10.4.6 Ductility

Ductility is an important property that can be used to determine if a material can be formed. It determines the capacity of the material to deform before it breaks.

A material that cannot sustain deformation is called a brittle material. Ductility is measured as either percent elongation, $\%e$, in the material or percent reduction in area, $\%AR$:

$$e = \frac{L_f - L_o}{L_o} \times 100$$

or

$$AR = \frac{A_o - A_f}{A_o} \times 100$$

10.4.7 Fatigue Properties

When the component is subjected to cyclic loading, fatigue properties become essential. In this case, the failure will be introduced at a stress lower than the yield stress. A correlation that has been used to estimate the fatigue limit is

Fatigue limit $= 0.5\,S_{\text{uts}}$

where S_{uts} is the ultimate tensile strength.

10.4.8 Impact Properties

The impact property of a material is its resistance to fracture under sudden impact.

10.4.9 Hardness

Hardness is the measure of the material's resistance to indentation. Hardness can be measured in three different tests: Brinell (uses balls to indent in the surface), Vickers (uses pyramids to indent the surface), and Rockwell (measures the depth of indentation). The Brinell hardness number (BHN) can be obtained as

$$BHN = \frac{2P}{D\left[D - \sqrt{D^2 - d^2}\right]}$$

where P is the load, D is the diameter of the indenter, and d is the measure of the diameter of indentation. The BHN is used to estimate the ultimate tensile strength

$$\left(SP_{uts} = \frac{P_{max}}{A_o}\right)$$

using

$$S_{uts} = 3.45 BHN$$

in MPa

$$S_{uts} = 0.5 BHN$$

in Ksi, where K is kips.

10.5 BILL OF MATERIALS

Once the materials have been selected, they should be represented in a *bill of materials*. This is an index of the parts that were used in the product. A typical bill of materials is shown in Table 10.2.

The following information should be included in a bill of material.

1. *The item number:* This is a key to the components on the assembly drawing.
2. *The part number:* This is a number used throughout the purchasing, manufacturing, and assembly system to identify the component. The item number is a specific index to the assembly drawing; the part number is an index to the company system.
3. *The quantity needed in the assembly.*
4. *The name and description of the component.*

TABLE 10.2 Bill of Materials

Item	Part	Quantity	Name	Material	Source
1	G-9042-1	1	Governer body	Cast aluminum	Lowe's
2	G-9138-3	1	Governer flange	Cast aluminum	Lowe's
9	X-1784	4	Governer bolt	Plated steel	Fred's Fine Foundry

5. *The material from which the component is made.*
6. *The source of the component.*
7. *The cost of the individual component:* This part will be kept for the design team.

LAB 10:   Material Selection Tutorial[1]

This lab describes an example of selecting optimal materials for a car key. Figure L10.1 shows a schematic diagram of a car-key structure. As seen from Figure L10.1, a car key has two main components: **grip** and **shaft**. The grip should be soft and cool to the touch, ensuring the comfort of a driver. On the other hand, the shaft should not break while twisted. At the same time, the shaft should be a good conductor of electricity so that it helps close an electrical circuit when starting the ignition.

GRIP The indices for the material selection of a grip are explained, as follows.

A material feels soft (to the touch) if its hardness is low and is flexible. As such, an index of softness can be formulated as

$$S = C(HE) \qquad (1)$$

In Equation (1), H is the hardness, E is the modulus of elasticity, C is a constant, and S is an index. To get a soft material, S has to be minimized. The hardness of material and its strength are directly correlated as

$$H = C'\sigma_y \qquad (2)$$

In Equation (2), σ_y is the yield strength of a material. Rearrangement of Equations (1) and (2) yields:

$$S = CC'(\sigma_y E) \qquad (3)$$

This implies a material index M_s that has to be maximized to get a soft material, as

$$M_s = \frac{1}{\sigma_y E} \qquad (4)$$

What about the 'coolness?' Suppose that Q amount of heat has been applied to V volume of material. This will raise the temperature of the material. This implies that

$$Q = mC_p\Delta T = \rho V C_p\Delta T \qquad (5)$$

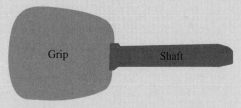

Figure L10.1   A schematic diagram of a key.

[1] AMM Sharif Ullah, Dr. Engineering, Department of Mechanical Engineering, Kitami Institute of Technology, Japan.

In Equation (5), m is the mass, C_p is the heat capacity, ρ is the density, and ΔT is the rise of heat temperature. Rearranging the variables in Equation (5) yields:

$$\Delta T = \left(\frac{Q}{V}\right)\frac{1}{\rho C_p} \tag{6}$$

To achieve 'coolness', ΔT has to be minimized. This implies a material index M_C that has to be maximized:

$$M_C = \rho C_p \tag{7}$$

To get an optimal material for the grip, the following formulation can be used:

$$M_{\text{optimal}} = w_S M'_S + W_C M'_C \tag{8}$$

where

$$W_S + W_C = 1$$

$$M'_S = \frac{M_S - \min(M_S)}{\max(M_S) - \min(M_S)}$$

$$M'_C = \frac{M_C - \min(M_C)}{\max(M_C) - \min(M_C)}$$

Figure L10.2 shows an MS Excel™-based system for selecting optimal materials for the grip of a car key. As seen from Figure L10.2, among five materials (aluminum alloy, silica glass,

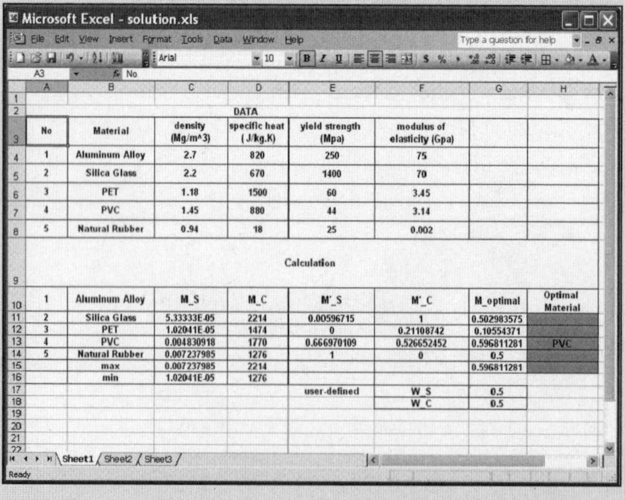
Microsoft Excel - solution.xls

DATA

No	Material	density (Mg/m^3)	specific heat (J/kg.K)	yield strength (Mpa)	modulus of elasticity (Gpa)
1	Aluminum Alloy	2.7	820	250	75
2	Silica Glass	2.2	670	1400	70
3	PET	1.18	1500	60	3.45
4	PVC	1.45	880	44	3.14
5	Natural Rubber	0.94	18	25	0.002

Calculation

1	Aluminum Alloy	M_S	M_C	M'_S	M'_C	M_optimal	Optimal Material
2	Silica Glass	5.33333E-05	2214	0.00596715	1	0.502983575	
3	PET	1.02041E-05	1474	0	0.21108742	0.10554371	
4	PVC	0.004830918	1770	0.666970109	0.526652452	0.596811281	PVC
5	Natural Rubber	0.007237985	1276	1	0	0.5	
	max	0.007237985	2214			0.596811281	
	min	1.02041E-05	1276				
				user-defined	W_S	0.5	
					W_C	0.5	

Figure L10.2 Material selection for the grip of a car key.

PET, PVC, and natural rubber), PVC is the optimal material given where equal importance has been placed on softness and coolness. The solution might change if the weights w_S and w_C are redefined.

GROUP TASK Use the aforementioned procedure and find out the optimal material for the shaft of the same car key by referring to the first paragraph of this lab for the shafts desired attributes. ■

10.6 GEOMETRIC DIMENSIONING AND TOLERANCING

The misperception that CAD models contain all of the information needed to produce the product has led to the importance of dimensioning and tolerancing. For years, the value of geometric dimensioning and tolerancing (GD&T) has been highly regarded due to its ability to provide a clear and concise method of communicating tolerance requirements for parts and assemblies when needed.[2]

One of the most important features of design work is the selection of manufacturing tolerance so that a designed part fulfills its function and yet can be fabricated with a minimum of effort. It is important to realize the capabilities of the machines that will be used in the manufacturing and the time and effort to maintain a tolerance.

It must be emphasized that in order for the GD&T to be effectively used, the functional requirements of the product must be well formulated. Vague requirements may result in a product that will not function as intended. Similarly, if tolerances are not well detailed, this may also cause a part not to function as intended or make it too expensive. Figure 10.2 on the next page shows two samples of outputs for a cup in which (a) the dimensioning and tolerancing standards are not used, and (b) detailed tolerance requirements are stated. It is clear that the output in part (a) results in a nonfunctional part, while in part (b), it results in a functional part. The translations of the tolerancing symbols in part (b) are as follows.

1. The top must be in line with the bottom within 2 mm.
2. The cup must not rock; the bottom must be flat within 0.5 mm.
3. The cup must be transparent.
4. The cup must have a smooth surface within $-0.4\ \mu$m.
5. The volume is 0.5 to 0.6 L.
6. The side angle is approximately 20 degrees ± 1 mm uniform zone.

Designers may have the tendency to incorporate specifications that are more rigid than necessary when developing a product. Figure 10.3 on the next page shows the relation between the relative cost and the tolerance. As the tolerance requirement becomes smaller, the cost becomes higher. Because of the high cost of tight tolerance, designers need to look closely at their design and adjust functions or components such that the tolerance will not add cost to the product. For example, if a part such as a bearing calls for a tight tolerance (0.005) when installed inside a casing and may function with (0.03) tolerance when placed outside the casing, designers need to change the design to fit the bearing outside the casing.

---

[2]Murphy, M. presentation at the NSF workshop at Central Michigan University, 1999.

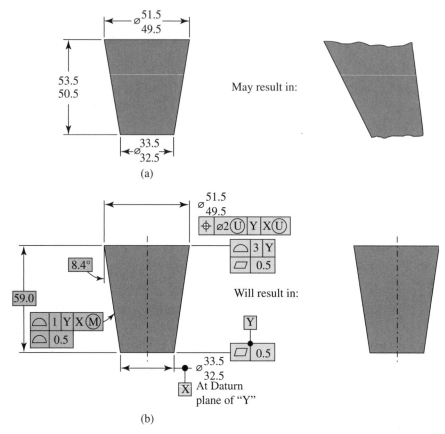

Figure 10.2   Comparison between lack of proper tolerancing and proper tolerancing.

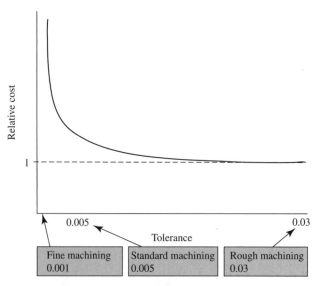

Figure 10.3   Tolerance versus cost.

LAB 11:   Geometric Dimensioning and Tolerancing

Procedure

This lab introduces you to geometric dimensioning and tolerancing.[3]

The GD&T Trainer demo CD-ROM can be obtained from Effective Training, Inc., and is based on the ASME Y14.5M code. Follow these instructions.

1. Launch the GT demo. There will be a few slides to explain the system requirements and what the program is all about.
2. Click on the course index.
3. Click on Concepts. Remember, this is just a demo version. About three modules will work in this demo.
4. Click on lesson number 3: Modifiers & Symbols. This module will introduce you to the geometric characteristics and symbols. The text that is marked red is clickable and will give you more information.
5. Proceed in this lesson. On page 10 there is a problem that you need to solve. Type that characteristic table as a Word document. You can hand draw the symbols if you can't use Word to reproduce them.
6. Go through the lesson and answer all the questions. Report all your answers in a Word document.
7. Once done with concepts, go back to the course index and click on Forms.
8. Then choose flatness.
9. Go through this tutorial and report your answers.
10. Once you have finished with flatness, go back and take the quiz located on the page for flatness. Report the quiz questions and answers.
11. Exit the demo and start Qr_demo From the subject index, choose position RFS/MMC and report your findings.

---

[3]The information is based on the ASME Y14.5M-1994 standard. The basis of this lab is obtained from The Effective Training Inc. demo. ∎

LAB 12:   Use of Pro/MECHANICA® for Structural Analysis

Purpose

Once the materials have been selected and the initial dimensions and tolerances specified, detailed analysis of the product is advised. One way to do this is performing structural analysis using the finite element method. This lab will introduce you to the use of Pro/MECHANICA® as an analysis tool for structural components. Pro/MECHANICA is a tool available for use in conjunction with the Pro/ENGINEER® software. Pro/ENGINEER® (Pro/E) is a solid-modeling software that also can be used to perform virtual prototyping.

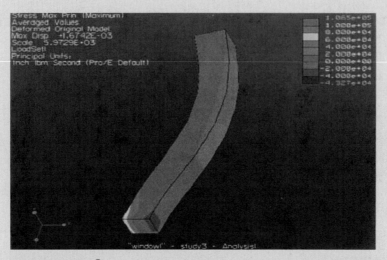

Pro/MECHANICA® from Parametric Technology Corporation is a solid modeling software designed to help the engineer design and test products using virtual prototyping.

Supported Beam

A 2-D simply supported beam by definition has a pin support at one end and a roller support at the other end. The essential feature of a pin support is that it restrains the beam from translating both vertically and horizontally but does not prevent rotation.

As a result, a pin joint is capable of developing a reactive force for both vertical and horizontal components, but there will be no moment reaction. A roller support prevents translation in the vertical direction but not in the horizontal direction. The roller also allows free rotation in the same manner as the standard pin joint.

Point and curve constraints on solid elements can cause singular stresses. The following procedure illustrates techniques for modeling a solid simply supported beam using rigid connections.

- Generate the part using Pro/E. The part is a square beam as shown in Figure L12.1
- Launch MECHANICA Application>MECHANICA. The default units are in (in-lbm-sec). For now assume the default units.
- Select **Structure** from the MECHANICA menu.
- Create datum planes at the ends of the beam. Select insert **Datum Plane**. From the menu shown in Figure L12.2, select **Parallel>Plane**. Select one of the end faces. Repeat for the other face.
- Add rigid connections on each end surface. Click on **Model>Idealizations> Rigid Connection>Create**, and pick one of the end surfaces. Repeat at the other end of the beam.
- Add constraints to each of the end planes. Click on **New Edge/curve constraints**, and pick one of the end surfaces. Fix all Translations as shown in Figure L12.3. Repeat for the other edge; this time leave the Z-Translation free.

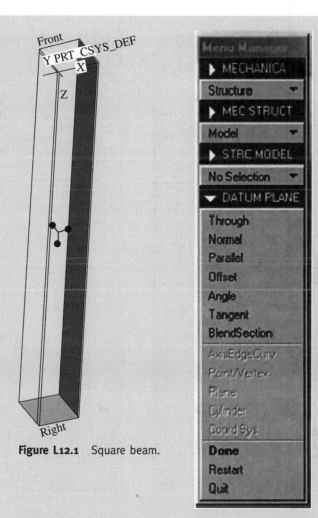

Figure L12.1   Square beam.

Figure L12.2   Menu manager.

- Apply Loads (see Figure L12.4). Click on **Model>Loads>New**.
- Apply Material Properties **Model>Material**.
- Check Model for errors **Model>Check Model**.
- Select the analysis type. Click on **Model>Analysis>Static>New>Close**.
- Create **Design Study Model>Design Study**.
- Run the analysis. Check the setting. Make sure you know where the files are going to be saved.
- Check the results. Create a results window. Check for the displacement (Figure L12.5) and stresses (Figure L12.6).

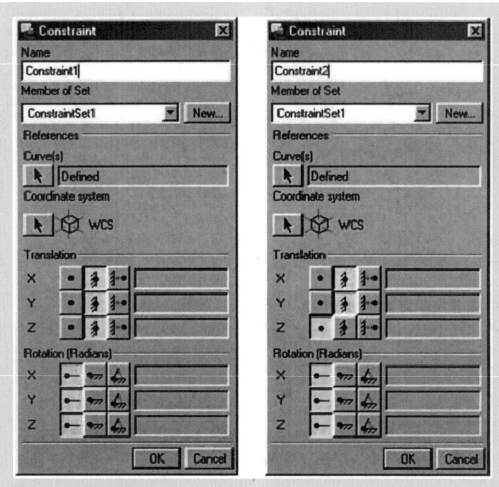

Figure L12.3 Constraints.

1. Use Pro/MECHANICA to solve for the stress displacement for a shaft of diameter = 2.0 and length = 10.00. A pin on one end and a roller on the other end support the shaft. Solve for the following loads:
   a. Material is steel and the load is uniformly distributed and equal to 10,000.
   b. Material is Al and the load is uniformly distributed and equal to 10,000.
   c. The load is concentrated at the middle and is equal to 100,000 (for both materials).
   d. The load is concentrated at a point 1/3 from the pin joint and is equal to 100,000 (for both materials).

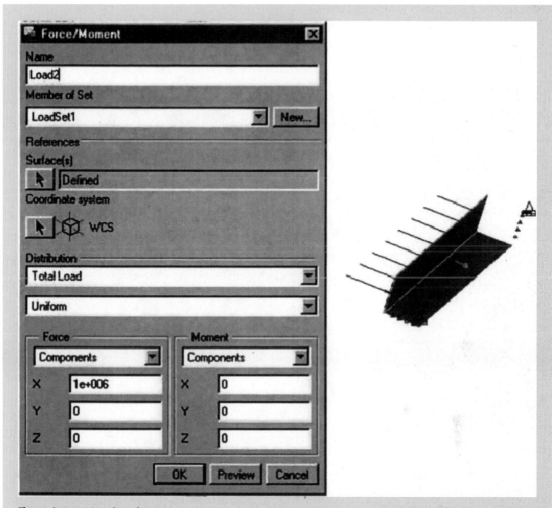

Figure L12.4 Load on beam.

    e. The load is concentrated at a point 2/3 from the pin joint and is equal to 100,000 (for both materials).

    f. Compare the results.

2. Using the beam in the example, change the dimensions of the beam to ($10 \times 2 \times 100$). Apply a uniform load of 10,000. (Material is bronze.)

    a. Find the displacement and stresses under this condition.

    b. Assume the analysis is 2-D (use the model type option with plane stress). Find the displacement and stresses; compare with part (a).

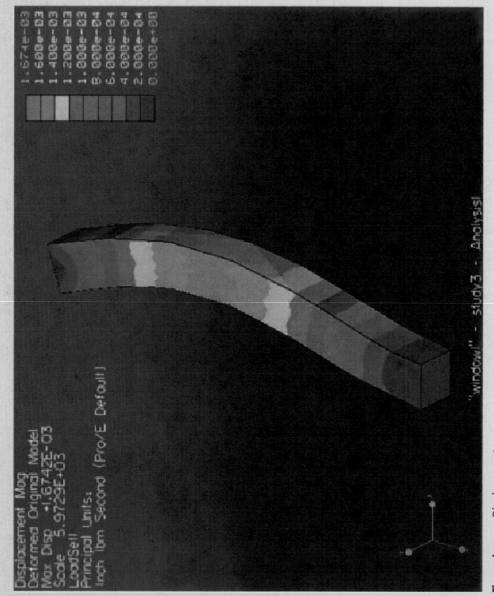

Displacement Mag
Deformed Original Model
Max Disp +1.6742E-03
Scale 5.9729E+03
LoadSet1
Principal Units:
Inch lbm Second (Pro/E Default)

1.674e-03
1.600e-03
1.400e-03
1.200e-03
1.000e-03
8.000e-04
6.000e-04
4.000e-04
2.000e-04
0.000e+00

"window1" - study3 - Analysis1

Figure L12.5 Displacement.

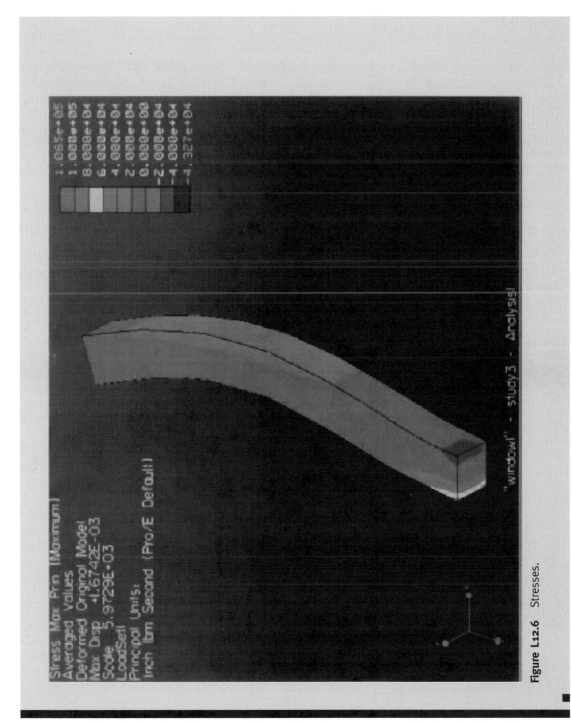

Figure L12.6 Stresses.

10.7 ANALYSIS EXAMPLE: MECHANICAL VEGETABLE HARVESTING MACHINE

In this example, a design team was asked to design and perform structural analysis on a mechanical vegetable collection and packaging system that is composed mainly of wood. The vegetable harvesting and packaging system must be capable of harvesting the vegetables without damaging a substantial amount of the crop. In addition, the design must not rely on manmade power in order to operate it, since it is to be used in the field. During the design process, the design team developed the function analysis shown in Figure 10.4. A morphological chart for the design is shown in Figure 10.5. Three different concepts were generated. Figure 10.6 on page 252 shows the three different concepts.

In the first concept, a horse pulls the yoke moving the vehicle forward. The front wheels turn a belt, which in turn provides torque on the blade shaft. Blades and sifters sort debris from the vegetables. Vegetables are then slung into the carriage compartment where an operator packages product.

The principle on which concept ll's mechanism operates is that vegetables are harvested very close to the soil surface. By taking advantage of a plow, the vegetables will be ripped free from the roots and travel up the plow to the conveyor system.

The only piece of metal found on the apparatus is found at the edge of the plow. This edge is reinforced with steel, since it will be in direct contact with the soil. The conveyor system is composed of two axles and a conveyor belt; the belt is composed of hemp (the strongest natural plant fiber). The conveyor system was used for sorting out the vegetables from the dirt and small stones—more or less like a sifter. The vegetables will stay on the conveyor system, and the soil will fall through the sorter. The rear axle drives the conveyor system via a three-to-one gear ratio from the rear axle to the driving axle of the conveyor system. This allows the conveyor to travel three times the speed of the tangential velocity of the rear axle. Once at the top of the conveyor system, the vegetables are pushed onto a rear-loading table, complete with guides. The rear axle is attached rigidly to the rear axle via a single wooden pin per wheel. The wheel has a carved out section where this pin lies. There is also a wedge that holds the wheel firmly against the pin to ensure a rigid connection.

The conveyor is driven via a friction belt located between the rear axle gears. The belt is arranged in a figure eight to allow the rotation induced upon the belt by the rear axle to be reversed so that the conveyor can travel in the correct direction, towards the back of the machine. The chassis is designed using a single piece of wood. Openings are included in the bottom of the chassis to allow soil or rocks to fall back to the earth. The main portion

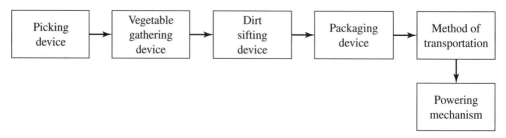

Figure 10.4   Function analysis for vegetable harvesting machine.

	Option 1	Option 2	Option 3	Option 4
Vegetable picking device	 Conveyor belt	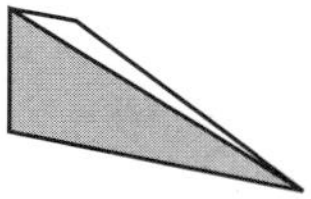 Triangular plow	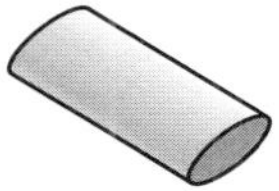 Tubular grabber	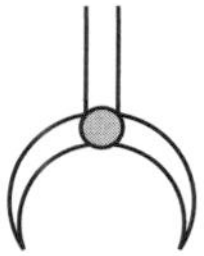 Mechanical picker
Vegetable placing device	Square mesh	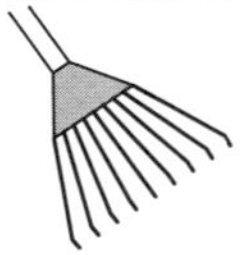 Rake	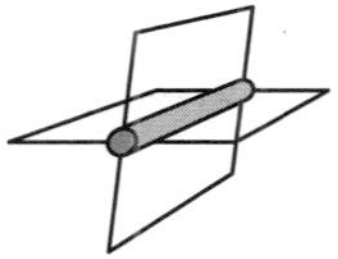 Rotating mover	 Force from vegetable accumulation
Dirt sifting device	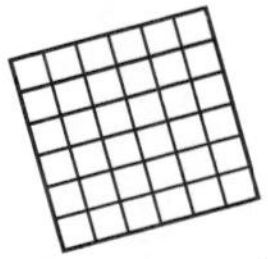	 Water from well	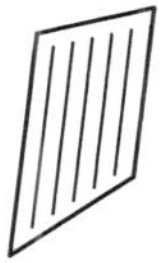 Slits in plow or carrier	
Packaging device		Track system	Sled	
Method of transportation				
Power source	Hand pushed	Horse drawn	Wind blown	Pedal driven

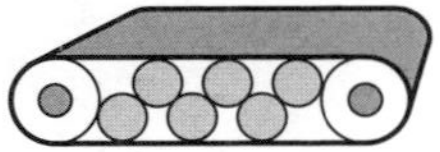

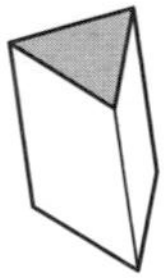

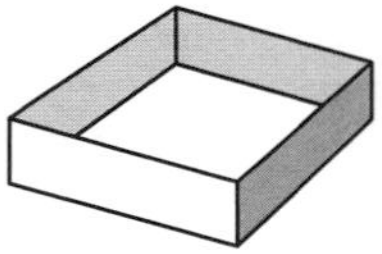

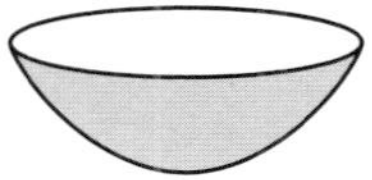

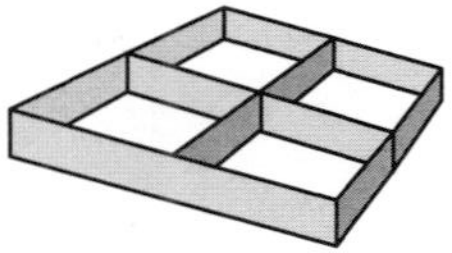

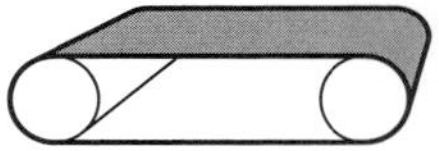

Figure 10.5 Morphological chart for vegetable harvesting machine.

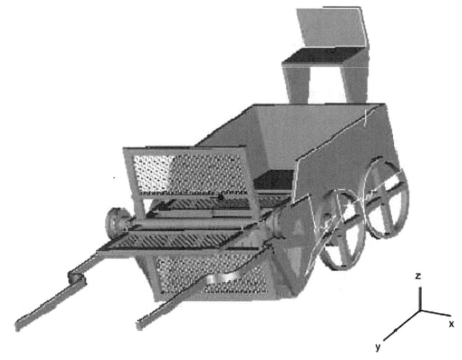

Figure 10.6a   Concept I for vegetable harvesting machine.

Figure 10.6b   Concept II for vegetable harvesting machine.

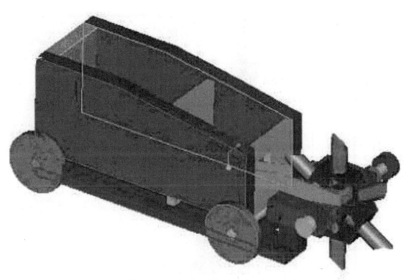

Figure 10.6c   Concept III for vegetable harvesting machine.

of the plow is made from the chassis, but is replaceable if severely damaged. The attachments where the horses are fastened are also made from the chassis. All of the wooden pieces are composed of red oak.

In concept III, a horse pulls the device in the *y* direction. The front wheels turn, making the collecting wheel rotate. The tubes cut the vegetables, dragging them against the cart. Once the vegetables are against cart, they are guided inside the pipes to the upper position. There is a hole at the bottom of each pipe where the vegetables exit. At this point, there is a slide that carries the vegetables to the back of the cart.

All three concepts were evaluated for their structural integrity. As an example, the structural analysis of the second concept will be presented here.

This analysis was done using Pro/MECHANICA software. Figure 10.7 on the next page shows the displacement and stress of the steel blade that is placed in the front of the plow. A load is applied to the edge of the blade. Figure 10.8 on the next page shows the load distribution on the chassis, and Figure 10.9 on page 255 shows the stress and displacement of the chassis due to these loads. Figure 10.10 on page 255 shows the loads on the rear shaft, and Figure 10.11 on page 256 shows the stresses and displacement due to these loads. Figure 10.12 on page 256 shows the loads on the wheel, while Figure 10.13 on page 257 shows the stresses and displacements due to these loads. The load results illustrated in Figures 10.7 through 10.12 show that the there would be no failure due to loading.

This type of analysis should be used as criteria when evaluating different designs. Designers should produce evaluation tables based on a part's performance under analysis. Designers could then change a part if the part has failed to fulfill its function.

An optimization for the concept based on the evaluation criteria, structural and functional analysis, and cost should be performed. In general, a mathematic model is developed to describe the different parameters influencing design. For example, a model to describe the cost based on the material choice, a model to describe the structural integrity, a model to describe the geometrical contribution to cost and structural integrity, and so on. These parameters are then optimized to find the alternative that will provide the best performance

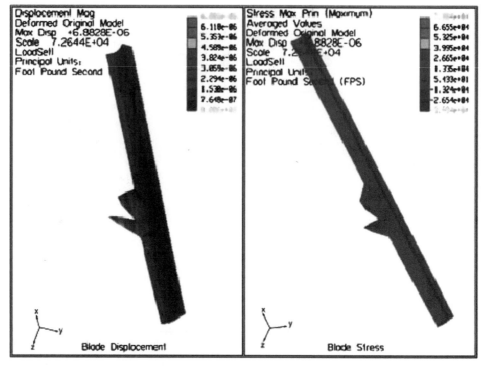

Figure 10.7 Displacement and stresses for the steel blade.

with the lowest cost. The optimization techniques are mathematical algorithms that have an active area of research.

Many researchers are studying the building of a real-time design engine where the alternatives are modeled using Pro/E or other solid modelers. An analysis engine then would evaluate the alternative for its performance criteria if a change is needed based on the analysis. Then a parameter or group of parameters will be changed and the analysis run one more time. This procedure continues until an optimized solution is obtained.

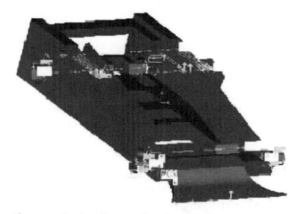

Figure 10.8 Loading on the chassis.

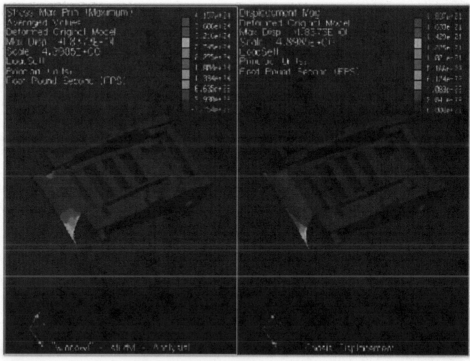

Figure 10.9   Displacement and stresses on chassis.

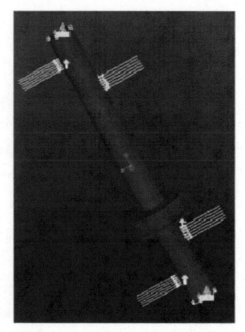

Figure 10.10   Loading on the rear shaft.

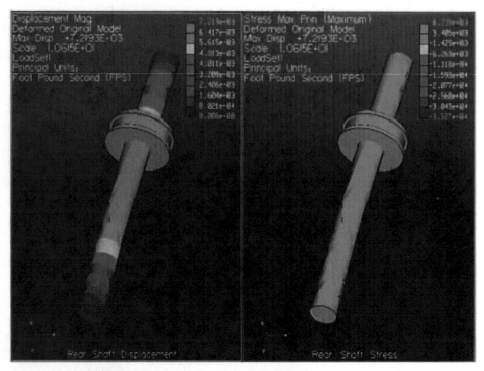

Figure 10.11 Displacement and stresses on rear shaft.

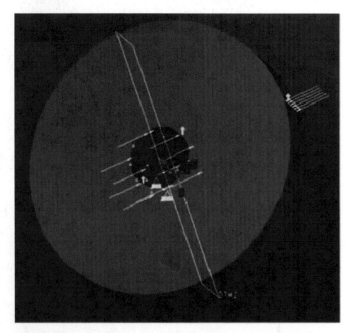

Figure 10.12 Loading on the large wheel.

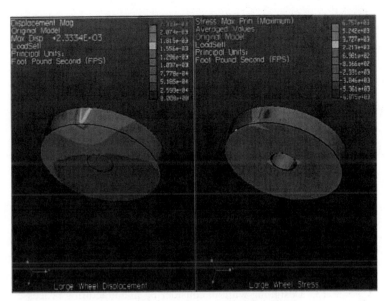

Figure 10.13   Displacement and stresses on large wheel.

10.8   COST ANALYSIS

Although it is important to generate a rough cost estimate as early as possible in the design process and then again to refine the cost estimate as soon as the embodiment phase is reached, it is only during the detailed design phase that a detailed cost analysis can be made. This is because it is only at this stage that the appropriate materials and manufacturing processes are identified as well as accurate dimensions and tolerancing specified. As the design is refined, the cost estimate is also refined until the final product is produced.

Cost evaluation includes, amongst other things, estimating the cost of building a plant or the cost of installing a process within a plant to produce a product or a line of products. It also involves estimating the cost of manufacturing a part based on a particular sequence of manufacturing steps.

In design, it becomes important to find techniques that help reduce cost. Some of the cost reduction techniques that can be used are as follows.

1. Introduce new processes.
2. Take advantage of knowledge gained as experience increases. To allow for standardization of parts, materials, and methods.
3. Employ a steady production rate, when feasible.
4. Utilize all production capacity.
5. Have product-specific factories or production lines.
6. Improve methods and processes to eliminate rework, reduce the work in process, and reduce inventory.

It is important to avoid factors that tend to increase the costs of the design process. Factors that increase costs are as follows.

1. Incomplete product design specifications.
2. Redesign due to failures.
3. Supplier delinquency.
4. Management and/or personnel changes.
5. Relocation of facilities.
6. Unmet deadlines.
7. A product that has become too complex.
8. Technology and/or processes that are not developed as well as they were thought to be.
9. Inadequate customer involvement.
10. Disregarding needs.
11. Not subjecting a new product idea to competitive evaluation.
12. Not demonstrating that a new design can function properly under realistic conditions.
13. Designing without considering manufacturing processes.
14. Not continually improving and optimizing the product.
15. Relying on inspection (last minute test).

10.9 COSTS CLASSIFICATIONS

Each company or organization develops its own bookkeeping methods. In this section, the different classification divisions of costs are presented. Cost estimation within a particular industrial or governmental organization follows a highly specialized and standardized procedure, particular to the organization.

1. *Nonrecurring-recurring:* There are two broad categories of costs:
   a. Nonrecurring costs: These are one-time costs, which are usually called capital costs, such as plant building and manufacturing equipment.
   b. Recurring costs: These costs are direct functions of the manufacturing operations.
2. *Fixed-variable costs:* Fixed costs are independent of the rate of production of goods; variable costs change with the production rate.
   a. Fixed costs include
      i. Investment costs
         - Depreciation on capital investment
         - Interest on capital investment
         - Property taxes
         - Insurance
      ii. Overhead costs include
         - Technical service (engineering)
         - Nontechnical service (office personnel, security, etc.)
         - General supplies
         - Rental of equipment

      iii. Management expenses
- Share of corporate executive staff
- Legal staff
- Share of corporate research and development staff

      iv. Selling expenses
- Sales force
- Delivery and warehouse costs
- Technical service staff

  b. Variable costs:
      i. Materials
      ii. Direct labor
      iii. Maintenance cost
      iv. Power and utilities
      v. Quality-control staff
      vi. Royalty payments
      vii. Packaging and storage cost
      viii. Scrap losses and spoilage

3. *Direct-Indirect:* A cost is called a direct cost when it can be directly assigned to a particular cost center, product line, or part. Indirect costs cannot be directly assigned to a product but must be spread over the entire factory.

  a. Direct costs include
      i. Material: Includes the expenses of all materials that are purchased for a product, including the expense of waste caused by scrap and spoilage.
      ii. Purchased parts: Components that are purchased from vendors and not fabricated inhouse.
      iii. Labor cost: Wages and benefits to the workforce needed to manufacture and assemble the product. This includes the employees' salaries as well as all fringe benefits, including medical insurance, retirement funds, and vacation times.
      iv. Tooling cost: All fixtures, molds, and other parts specifically manufactured or purchased for production of the product.

  b. Indirect costs include
      i. Overhead: Cost for administration, engineering, secretarial work, cleaning, utilities, leases of buildings, insurance, equipment, and other costs that occur day to day.
      ii. Selling expenses: marketing advertisements

10.10 COST ESTIMATE METHODS

The cost estimating procedure depends on the source of the components in the product. There are three possible options for obtaining the components.

1. Purchase finished components from a vendor.
2. Have a vendor produce components designed inhouse.
3. Manufacture components inhouse.

TABLE 10.3 The Make–Buy Decision

Reason to Make	Reason to Buy
Cheaper to make	Cheaper to buy
Company has experience making it	Production facilities are unavailable
Idle production capacity available	Avoid fluctuating or seasonal demand
Compatible and fits in production line	Inexperience with making process
Part is proprietary	Existence and availability of suppliers
Wish to avoid dependency on supplier	Maintain existing supplier
Part fragility requiring high packing	Higher reliability and quality
Transportation costs are high	

There are strong incentives to buy existing components from vendors. Cost is only one factor that determines whether a product or a component of a product will be made inhouse or bought from an outside supplier.

If the quantity to be purchased is large enough, most vendors will work with the product design and modify existing components to meet the needs of the new product. If existing components or modified components are not available off the shelf, then they must be produced. In this case, a decision must be made regarding whether they should be produced by a vendor or made inhouse. This buy–make decision is based on the cost of the component involved, as well as the capitalization of equipment and the investment in manufacturing personnel. Table 10.3 lists factors effecting the make-buy decision.

When manufacturing an element, the methods used to develop cost evaluation fall into three categories:

1. *Methods engineering (industrial engineering approach):* The separate elements of work are identified in detail and summed into total cost per part. Machined components are manufactured by removing portions of the material that are not wanted. Thus, the costs for machining are primarily dependent on the cost and shape of the stock material, the amount and shape of the material that needs to be removed, and how accurately it must be removed. An example for the production of a simple fitting from a steel forging is shown in Table 10.4.

2. *Analogy:* Future costs of a project or design are based on past costs of a similar project or design, with due allowance for cost escalation and size difference. This method, therefore, requires a backlog of experience or published cost data.

3. *Statistical approach:* Techniques such as regression analysis are used to establish relations between system costs and initial parameters of the system: weight, speed, power, etc. For example, the cost of developing a turbofan aircraft might be given by

$$C = 0.13937 \times 10.7435 \times 20.0775$$

where C is in millions of dollars, $\times 1$ is the maximum engine thrust in pounds, and $\times 2$ is the number of engines produced.

TABLE 10.4 Sample Production/Operation Cost Table of a Simple Fitting from a Steel Forging

Operations	Material	Labor	Overhead	Total
Steel forging	37.00			37.00
Set-up on milling machine		0.2	0.8	1.00
Mill edges		0.65	2.6	3.25
Set-up on drill press		0.35	1.56	1.91
Drill 8 holes		0.9	4.05	4.95
Clean and paint		0.3	0.9	1.2
Total	37.00	2.40	9.91	49.31

10.11 LABOR COSTS

The most accurate method of determining labor cost is to set up a staffing table to find the actual number of people required to run the process line. Be sure to account for standby personnel. Cost of operating labor includes fringe benefits. As an order of magnitude estimate of labor costs,

$$\frac{\text{Operating person hour}}{\text{Tons of product}} = K\frac{\text{Numbers of process steps}}{(\text{tons/day})^{0.76}}$$

where

$K = 23$ for a batch operation

$K = 17$ for average labor requirements

$K = 10$ for an automated process

10.12 PRODUCT PRICING

Pricing can be defined as the technique used for arriving at a selling price for a product or service that will interest the customer while simultaneously returning the greatest profit. Different methods are available to assist in arriving at the optimum price for a product. One of these methods is known as the break-even chart.

10.12.1 Break-Even Chart

The break-even chart (Figure 10.14 on the next page) is designed to graphically show the profits and losses on the selling price and manufacturing costs of the product. The manufacturing costs of a product include.

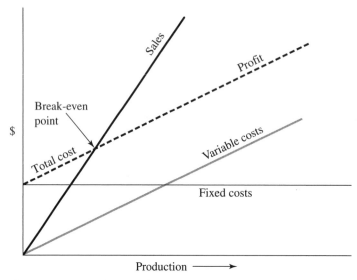

Figure 10.14   Break-even chart.

1. Material: Credit should be accounted for recycled scrap and by-products
2. Purchased parts
3. Labor
4. Tooling
5. Overhead

Material, purchased parts, and labor are variable costs that depend on the number of goods produced while tooling. Overhead is a fixed cost, regardless of the number of goods produced. The fact that the variable costs depend on the rate or volume of production while fixed costs do not leads to the idea of a break-even point. Determination of the production lot size needed to exceed the break-even point and produce a profit is an important consideration. To draw a break-even point chart, use the following steps.

Step 1.   *Determine the variable costs per product.*
Variable costs depend on the number of products and include
a. Material
b. Purchased parts
c. Labor

Step 2.   *Determine the fixed cost.*
Fixed cost does not depend on the volume or the number of products.

Step 3.   *Draw, on the chart, the total cost as a function of the number of products.*

The total cost = fixed cost + variable cost

Step 4. *Determine a sale price of the product.*
        Draw the sales as a function of the number of products.

Step 5. *The intersection of the sales line with the total price is the break-even point.*

10.12.2 Linear Programming

Linear programming is a tool that is used to represent a situation, in which an optimum goal is sought, such as to maximize profit and minimize cost. Mathematically, it involves finding a solution to a system of simultaneous linear equations and linear inequalities, which is optimized in linear form. An illustrative example is listed here for demonstration of this technique.

A manufacturer of desk staplers and other stationary items produces two different stapler models (see Table 10.5). One is manual in operation, retails at $12.50, and returns $2.00 to profit. The other is automatic and sells for $31.25, returning $5.00 to profit. The manufacturer is concerned with how to utilize his production facilities in order to maximize profits. One constraint placed on the maximizing of profits can be written as follows:

$$P = 2Q_M + 5Q_A$$

where P is profit function, Q_M is the quantity of manual staplers, and Q_A is the quantity of automatic staplers.

Assume that the manufacturing operation is such that labor skills and plant facilities are completely interchangeable; thus, varying quantities of manual and automatic staplers may be produced on the same day. Also, assume that the present demand is such that the company can sell as many as it can produce. There are certain production restraints. Most of the major components are prepared on separate assembly lines, prior to moving to the final assembly area. There is a sufficient inventory of components for both staplers to satisfy daily production, with the exception of solenoids for the automatic model, which has a daily production capacity of 200. This constraint may be written as

$$Q_A \le 200$$

TABLE 10.5 Stapler Models

	Manual	**Automatic**
Selling price	31.25	12.5
Profit	$2	$5
Constraints		power supply 200 per day
Labor	18 person-minutes	54 person-minutes
Adjustment	3 person-minutes	5.4 person-minutes
Inspection	1 person-minute	1.5 person-minutes

Among other duties of employees, fabrication and assembly requires 18 person-minutes for each manual model and 54 person-minutes for each automatic stapler. The plant employs 45 people for this task, on an 8-hour shift. Since there is only one shift, the capacity is $8 \times 45 \times 60$, or 21,600 person-minutes per day. This may be defined mathematically by

$$18Q_M + 54Q_A \leq 21{,}600$$

The final adjustment in the manual stapler requires 3 person-minutes on the assembly line, whereas the automatic model requires 5.4 person-minutes. This phase requires 6 people, and the daily capacity is $8 \times 6 \times 60$ or 2880 person-minutes. This constraint is expressed as

$$3Q_M + 5.4Q_A \leq 2880$$

Final inspection for workability requires 1 person-minute for manual models and 1.5 person-minutes for the automatic model. There are 4 half-time inspectors assigned to this task; the inspection capacity is $8 \times 2 \times 60$ or 960 person-minutes. The inspection constraint is

$$Q_M + 1.5Q_A \leq 960$$

The constraints equations are as follows:

$$P = 2Q_M + 5Q_A$$
$$Q_A \leq 200$$
$$18Q_M + 54Q_A \leq 21{,}600$$
$$3Q_M + 5.4Q_A \leq 2880$$
$$Q_M + 1.5Q_A \leq 960$$

Plotting these equations yields Figures 10.15 and 10.16.

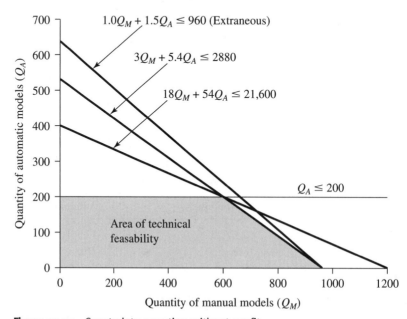

Figure 10.15   Constraints equation without profit.

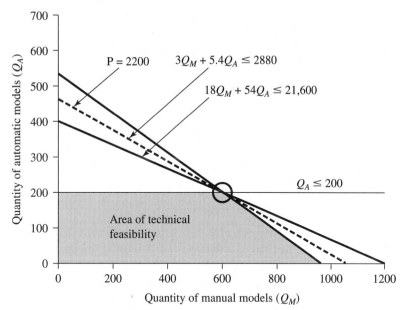

Figure 10.16 Constraints equation with profit.

Linear programming is used to find the isoprofit line (profit is the same along the line) that is farthest away yet still contains at least one point that is technically feasible.

A combination of break-even and linear programming approaches can be used to set a price for a product. The linear programming approach can be used to determine the optimum number of goods; then that number may be used on the break-even chart to determine the selling price. The actual setting of a price resides at the heart of business practice, but describing it briefly is not easy. A key issue in establishing a price involves the reaction of the competition and the customer. When competition is fierce, the sales force may push hard for price reductions, claiming that they can make up the lost revenues by increasing the volume. Special offers, such as promotional discounts, premiums, extras, trade-ins, volume discounts, and trade-in allowances, are often used.

10.13 PROBLEMS

10.13.1 Team Activities

The objective of this exercise is to use computer programming to write a program that can find the cost of different alternatives for your design project. Manufacturing cost is an important factor in evaluating the different alternatives. Some alternatives will be eliminated because of their high manufacturing cost.

Outline

In engineering design projects, students are expected to build a device by the end of a limited time period. With a limited amount of money, students need to figure out the manufacturing costs of different designs before they actually start building these devices. A manufacturing cost is a key determinant of the economic success of a product. When computing the manufacturing cost, one can divide the cost into two broad categories: (1) fixed costs and (2) variable costs. Fixed costs are incurred regardless of the number of products that are manufactured, whereas variable costs depend on the amount of goods that are produced. In student design projects, students are not expected to buy major tools or rent warehouses for production; hence the fixed costs and variable costs in this case may use the following components:

1. Fixed Costs
   a. *Tooling cost:* Tooling costs can include buying or renting. Renting depends on the charge per unit time, regardless of the number of units produced. Buying in this case will be for this project only, and it will be assumed that it is paid for in full (i.e., no interest and no depreciation).
   b. *Setup cost:* This is the cost required to prepare equipment for a production, and it is fixed regardless of the number of units produced.
2. Variable Costs
   a. *Material cost:* See Figure 10.17.
   b. *Processing cost:* Varies with the type of manufacturing equipment used, and includes charges for both machine time and labor.

Programming Requirements

The program should be readable; all items from the terminal are sorted into "ready" and "need-manufacturing" categories. The information should be fed by the user. Ready items need prices from the vendor, while items that need manufacturing require the following calculation:

1. User inputs the following information:
   a. Mass of the material (see provided table)
   b. Total number of units of this type
   c. Setup cost
   d. Tooling cost
   e. Processing cost
2. The computer calculates the manufacturing cost for each element.

The output should be in the format shown in Table 10.6.

10.13.2   Individual Activities

1. Name three different cost classifications, and state the differences among them.

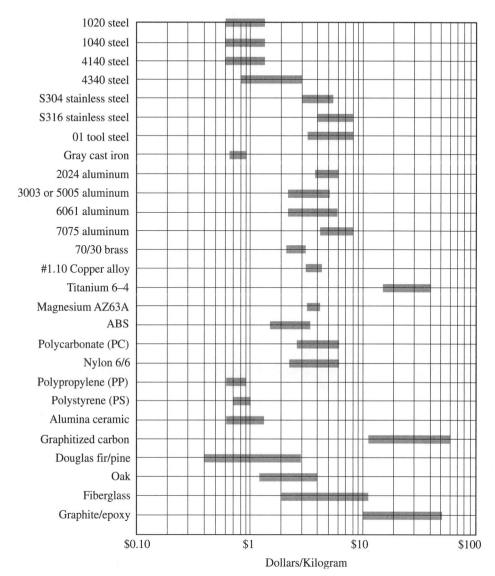

Figure 10.17 Material cost. Range of costs for common engineering materials. Price ranges shown correspond to various grades and forms of each material, purchased in bulk quantities (1994 prices). (Adapted from David G. Ullman, *The Mechanical Design Process*, McGraw-Hill, New York, 1992.)

2. What are the three cost estimate methods, and what are the differences among them? Which method is more appropriate for your design project and why?

3. What is the break-even chart?

4. Using the break-even method, find break-even points if
   a. The fixed cost is $12,000.
   b. The variable cost is $17.95 per unit.
   c. The sale price is $29.50 per unit.

5. What would be the break-even in the previous question if the fixed cost is the same but the variable cost becomes $29.50 and the selling price becomes $37.50?

6. The following data is for a company that produces washers and dryers.

   - Dryers retail for $198 and contribute $15 to profit.
   - Washers retail for $499.95 and contribute $45 for profit.
   - The washer blade is limited in production capacity to 50 blades, while all other components have no limits.
   - Chassis assembly requires 6 person-hours for each dryer set and 18 person-hours for each washer. The plant employs 225 workers for an 8-hour shift to perform chassis assembly operations.
   - A dryer requires 1 person-hour on the assembly line, a washer set 1.6 person-hours. There are 30 people on a single 8-hour shift assigned to assembly.
   - Final inspection requires 0.5 person-hour for a dryer and 2.0 person-hours for washer. The plant employs 20 full-time inspectors and one part-time employee for 2 hours per day.
     a. Write a linear programming model.
     b. What is the optimum number of dryers and washers?
     c. Calculate the maximum profit.
     d. The linear programming and break-even approaches are used to find the selling price for both the dryer and washer, based on $10,000 fixed cost for both dryer and washer with a variable cost of $160 for the dryer and $330 for the washer. Determine how many days will be required until the company starts making a profit for the washers and the dryers. Use $198 and $500 as selling prices for the dryer and the washer, respectively.

Material Selection

1. In two opposite columns, list five uses of metals that are not appropriate for plastics and five uses of plastics not appropriate for metals.

2. From the table of mechanical properties (see Table 10.6) of selected materials, list the differences between pure metals and metal alloys.

3. Based on the table, compare between metals-alloys/ceramics/plastics in terms of weight, strength, stiffness, and thermal expansion.

4. What is the difference between forging, casting, and welding?

5. What is meant by each of the following terms?
   (a) Elastic limit
   (b) Yield point
   (c) Ultimate strength
   (d) Stiffness

6. What is the major requirement of a material if minimum deflection is required? Give an example.

7. What is meant by ductility? Suggest a method to measure ductility.

TABLE 10.6 Material Properties

Material	Density lb/in^3	ν Poisson's Ratio	E 10^4 psi	σ_y 10^3 psi	σ_{ult} 10^3 psi	Melting Temperature °C	Thermal Conductivity lb/sec°F
Pure Metals							
Berylium	0.066		44	55	90	2340	19.9
Copper	0.32	0.33	17	10	32	1980	51
Lead	0.41	0.43	2	2	2.5	621	4.5
Nickel	0.32		30	20	70	2625	7.9
Tungsten	0.70		50		300	6092	26.2
Aluminum	0.1	0.33	10	3.5	11	1200	29.0
Alloys							
Aluminum 2024-T4	0.1	0.33	10.6	44	60	1075	15.8
Brass	0.31	0.35	15	60	74	1710	14.9
Cast iron (25T)	0.26	0.2	13	24	120	2150	5.8
Steel: 0.2% C							
Hot Rolled	0.283	0.27	30	40	70	2760	6.5
Cold Rolled	0.283	0.27	30	65	80	2760	6.5
Stainless Steel							
Type 302 C.R.	0.286	0.3	29	100	140	2575	1.9
Ceramics							
Crystalline Glass	0.09	0.25	12.5	20		2280	0.24
Fused Silica Glass	0.08	0.17	10.5			2880	0.17
Plastics							
Cellulose Acetate	0.047	0.4	0.25	5	20		0.032
Nylon	0.041	0.4	0.41	8	13		0.03
Epoxy	0.04		0.65	7	30		0.10

10.14 Selected Bibliography

DHILLON, B. S. *Engineering Design: A Modern Approach*. Toronto: Irwin, 1995.

DIETER, G. *Engineering Design*. New York: McGraw-Hill, 1983.

DYM, C. L. *Engineering Design: A Synthesis of Views*. Cambridge, UK: Cambridge University Press, 1994.

HILL, P. H. *The Science of Engineering Design*. New York: McGraw-Hill, 1983.

PUGH, S. *Total Design*. Reading, MA: Addison-Wesley, 1990.

RAY, M. S. *Elements of Engineering Design*. Englewood Cliffs, NJ: Prentice Hall, 1985.

RADCLIFFE, D. F. and LEE, T. Y. "Design Methods used by Undergraduate Students." *Design Studies*, Vol. 10, No. 4, pp. 199–207, 1989.

SUH, N. P. *The Principles of Design*. New York: Oxford University Press, 1990.

ULMAN, D. G. *The Mechanical Design Process*. New York: McGraw-Hill, 1992

ULRICH, K. T. and EPPINGER, S. D. *Product Design and Development*. New York: McGraw-Hill, 1995.

VIDOSIC, J. P. *Elements of Engineering Design*. New York: The Ronald Press Co., 1969.

WALTON, J. *Engineering Design: From Art to Practice*. New York: West Publishing Company, 1991.

CHAPTER · **11**

Selection of Design Projects

'Practice makes perfect!' In this chapter, we will present a selection of design projects, including one for this shopping cart.

11.1 DESIGN PROJECT RULES

This chapter introduces sample design projects. The following rules apply to all design projects. Specific rules will also be listed for each project.

1. Students will apply a systematic design process to build the prototype.

2. Students will develop an objective tree to clarify the need statement and prioritize the objectives laid out in the problem statement. Students may add features that may not be listed clearly in the problem statement but will give an additional advantage to the proposed design. The objective tree is a mechanism to organize the objectives or design criteria in a tree fashion and combine different objectives under a higher objective.

3. Students will conduct a market analysis. In the market analysis students will conduct surveys to test and collect the customers wishes. Students will also conduct a literature search of all industry using the Web and evaluate the future market. A patent search should also be conducted to prevent preoperative patent violations. The large market players should be identified. Study of the industry trends in the last 5 years should be conducted.

4. Students should develop a specification table. The specification table should recognize which of the specifications is a demand and which is a wish. The demands are laid out in the problem statement, while the wishes are obtained from the objective tree. A house of quality can be developed to enable students to better understand the objectives they have set up.

5. Students should develop a functional analysis of the project. In the function analysis students will lay down the functions that are involved in a working model.

6. Students will develop a morphological chart that includes all functions and mechanisms able to perform that function.

7. Students will generate several design alternatives by combining the mechanisms in the morphological chart in accordance with the function layout.

8. Students will evaluate the different alternatives by using the Pugh method to identify the best design.

9. Students will generate a detailed drawing with assembly of their design. Solid modeling will be utilized before a prototype is generated. Students will optimize their design and modify their design on Pro/E solid drawings.

10. Students will build a mock-up of the design to check the conceivability and the manufacturability of the different components.

11. Students will generate a bill of material and identify the vendors, cost, and different part alternatives.

12. Students will calculate the cost and use the break-even point to find the selling price of their product.

13. Students will use the machine shop to build their design project.

14. Students will be trained in the different design tools. Students will use the Gantt, CPM, or PERT methods for scheduling the different events in the design process.

15. Students will be introduced to the different manufacturing processes during the course of the design.

16. The device must be a stand-alone unit. No assistance from the operator will be allowed after actuation.

17. Any device that is available through a retailer will be disqualified. The device must be designed and constructed from the component level by team members.

18. Groups will be given a 2-minute warning before they compete. They must report to the competition site within this time or they will be disqualified.

19. Judges and/or instructors will disqualify any device that appears to be a safety hazard.

20. The total cost of the device should not exceed the given budget ($200). A tolerance of 10% may be allowed after discussion with the instructor.

21. A complete breakdown of all costs must be included in the report. Receipts must be supplied

22. Teams are to comprise four members, to be selected and decided on by the students.

23. The Rhomberg method will be used by students to evaluate each team member's contribution. Student evaluations of team members will be kept confidential and are to be turned in weekly. These evaluations will be considered for grading.

24. The first phase includes the objective tree, function tree, market analysis, ideas, concepts sketches, schedules, and material selection.

25. The second phase includes evaluation, optimization, analysis, modeling, prototype/mock-up, and marketing.

26. Teams may choose to exceed the budget by 10% (the overall cost should never exceed $220).

27. The technical report and marketing brochures must be turned in.

28. All devices will be collected and impounded two days before the competition (i.e., students will not be allowed to take the devices away until the completion of all presentations).

29. The top three projects will be showcased by the department. Commendations and prizes will be awarded. Judging will be based on the following criteria:
    a. Performance
    b. Appearance
    c. Safety
    d. Other general engineering principles

30. The judges'/instructor's decisions at all times will be final. Appeals by students will be heard, but once a decision is reached it will be considered as final and no further appeals will be tolerated.

31. Students are reminded of the general rules and honor codes of their respective institutions.

32. Students are encouraged to be creative and not restrict themselves to ideas that are currently available. This exercise should be a positive learning experience for

the students; competitiveness is encouraged. However, the spirit of fairness must always prevail, and students are requested to abide by the rules and decisions of the judges and instructor.

33. The instructor reserves the right to change/add rules to fit student learning needs. Students will be informed of the changes when they occur.

11.2 ALUMINUM CAN CRUSHER

11.2.1 Objective

Design and build a device/machine that will crush aluminum cans. The device must be fully automatic (i.e., all the operator needs to do is load cans into the device). The device should switch on automatically, crush the can automatically, eject the crushed can automatically, and switch off automatically (unless more cans are loaded).

11.2.2 Specifications

- The device must have a continuous can feed mechanism.
- Cans should be in good condition when supplied to the device (i.e., not dented, pressed, or slightly twisted).
- The can must be crushed to one-fifth of its original volume.
- The maximum dimensions of the device are not to exceed $20 \times 20 \times 10$ cm.
- Performance will be based on the number of cans crushed in one minute.
- Elementary school children (kindergarten up) must be able to operate the device safely.
- The device must be a stand-alone unit. No assistance from the operator will be allowed after actuation.
- Any device that is available through a retailer will be disqualified. The device must be designed and constructed from the component level by team members.

11.3 COIN SORTING CONTEST

11.3.1 Objective

Design and build a machine/device that will sort and separate different types of coins. A bag of assorted coins will be emptied into the device and the contestants will be given one minute to sort and separate as many coins as possible. Coins will be disbursed in a bag containing an assortment of U.S. coins (pennies, nickels, dimes, and quarters) and other foreign coins. The device must be able to sort out and separate only the U.S. coins. The volume of coins that the bag will contain is approximately 0.3% of a cubic meter. The coins will be deposited into the device at the start time. The number of coins sorted and separated will be counted at the end of time (60 seconds). Each incorrect coin separated will be penalized. Each coin that is physically damaged by the device will be penalized. Coins that remain in the internal body of the device at the end of time will not be counted. Each round will be run independently (i.e., the only opponent is the clock). Each team will compete twice on two different cycles, based on random selection. The better of the two performances will be considered in the contest.

11.3.2 Constraints

The total cost of the device should not exceed the given budget ($200). A tolerance of 10% may be allowed after discussions with the instructor. A complete break-down of all costs must be included in the report. Receipts must be supplied. The device must be a stand-alone unit. No assistance from the operator will be allowed after actuation. The operator will be allowed to power the device either manually or by any other means. Any device that is available through a retailer will be disqualified. The device must be designed and constructed from the component level by team members. If there are any questions, the instructor should be consulted for clarification.

11.4 MODEL (TOY) SOLAR CAR

11.4.1 Objective

Design and build a model solar car that can be used as a toy by children. The primary source of energy is solar.

The following are some methods to stimulate your thinking; you are not limited to these methods.

- Direct solar drive.
- Battery storage (Students should note that on the day of the competition, the battery must be completely discharged and they will be given half an hour to charge the battery, using solar energy.)
- Air or steam engine (Students will be allowed to use photocells, lenses, etc., for the purpose of heating the air, or steam generation.)

11.4.2 Specifications

- The car will be required to travel 30 m in a straight line. Extra credit will be given to cars that are remote controlled. The final competition will take place in a school parking lot.
- Performance will be based on the car that travels the distance of 30 m in the least amount of time.
- The car must travel a distance of at least 3 m or it will be considered damaged. Damaged cars will get a failing grade in the performance section of the competition.
- The dimensions of the car are limited to the following maximum values:
  Length: 30 cm
  Width: 15 cm
  Height: 15 cm

- The car must be safe for children to use.
- Any car that is available through a retailer will be disqualified. The car must be designed and constructed from the component level by team members.

11.5 WORKSHOP TRAINING KIT

The Stirling engine kit that is used for a Mechanical Engineering Tools class is becoming more expensive and difficult to obtain on time. The goal of using that kit is to provide students with hands-on experience using the machine shop and at the same time produce a machine that is able to convert heat into work. This project calls for students to design another kit that can be designed and manufactured at the College of Engineering and serve the same goal of providing hands-on experience and building a machine that is able to convert one form of energy to another. Design and build a machine shop training kit that can be used for a Mechanical Engineering Tools class.

11.5.1 Specifications

- The kit must have teaching and training value to the students. The kit must be innovative and utilize an engineering principle to produce work (such as the first law of thermodynamics).
- The kit must have identified training values for machine shop tools.
- The dimensions of the kit when assembled are limited to the following maximum values: Length: 30 cm width: 15 cm and height: 15 cm.
- The kit, once assembled, must be safe to use for students in kindergarten through grade 12.
- Any kit that is available through a retailer will be disqualified. The kit must be designed and constructed from the component level by team members.

11.6 SHOPPING CARTS

As technology advances and new materials are synthesized, certain products are not modified and modernized. Among those are the shopping carts that are being used in grocery stores. As many of you may have observed, there is a tendency to conserve parking space by not designating a return cart area. Leaving carts in the parking lots may lead to serious accidents and car damage. Furthermore, many customers do not fill their carts when shopping; however, they do not like to carry baskets. Other customers like to sort products as they shop.

Design and build a new shopping cart that can be used primarily in grocery stores. The shopping cart should solve the common problems in the available carts.

11.6.1 Specifications

You must adhere to the following rules:

- Conduct surveys to measure customer needs for shopping carts.
- The shopping cart should be safe for human operation.

- The dimensions of the cart should match existing carts.
- The cart should be easy to operate by children of age 7 and older and senior citizens.
- The cart should have features that accommodate children while shopping.
- The cart should be able to accommodate large and small items.
- Any cart that is available through a retailer will be disqualified.

11.7 MECHANICAL VENTS

Most houses have vents that open and close manually without any central control. Cities across the country advise the use of such vents to save energy. In most cases, household occupants do not use the entire house at the same time; the tendency is to use certain rooms for a long time. For example, the family room and dining room may be used heavily, while the living room and kitchen are used at certain hours of the day. To cool or heat a room, the system needs to work to cool or heat the entire house. Energy saving can be enhanced if the vents of unused rooms are closed; this will push the hot/cold air to where it is needed most and reduce the load of the air conditioning system.

Design and build a remotely operated ventilation system that will be connected to an existing ducting system. The house layout will be given upon request.

11.7.1 Specifications

You must adhere to the following rules:

- The system must be safe and easily adapted to existing ventilation systems.
- The system must work for at least seven exits.
- The venting system must accommodate high and low operating temperatures.
- The system must be easy to operate by household members.
- Central units of operation are encouraged as well as remote controlled units.
- Any system that is available through a retailer will be disqualified. The system must be designed and constructed from the component level by team members.

11.8 ALL TERRAIN VEHICLE

The objective of this project is to design and build a model for an all terrain vehicle (ATV). The size of the model should not exceed $15 \times 15 \times 15\,\text{cm}^3$. The power of the model vehicle is flexible, but safety must prevail. Success of this model design may lead to a new market for ATVs, such as

- A toy for children.
- A demonstration model for object collection on other planets.
- A demonstration model for law enforcement and military purposes.

The designed model must be able to handle slopes of 45 to 60 degrees, go over rocks that are the height of the wheels, and move in mud and dry dirt.

11.9 POCKET-SIZED UMBRELLA

As technology advances, the public is searching for more convenient products to replace existing products. In this project you are requested to design and build an umbrella that can fit in a normal pocket when folded. The following must be adhered to when designing the new umbrella:

- The umbrella must be able to cover the prospective customer from heavy rain.
- The response time to open and fold the umbrella must be reasonable and close to the response time of an existing full-size umbrella.
- The weight of the umbrella should be reasonable to fit in a pocket without damaging the pocket.

11.10 MODEL OF THERAPEUTIC WHEELCHAIR

In homes of the elderly and infirm and in medical therapy clinics, a nurse is requested to help residents/patients walk on a daily basis. Most of the residents and patients are unable to walk alone. They use standard wheelchairs most of the day. Because of cost factors, only one nurse can be assigned to each patient. The patient usually leaves the wheelchair and walks away from it with the help of the nurse. In this situation a problem may arise if the patient needs the wheelchair urgently while he and the accompanying nurse are away from it. The nurse cannot leave the patient unattended to bring the chair; neither can she carry the patient back to the wheelchair. You are required to design and build a prototype wheelchair that will provide the necessary solution for the nurse. The model should not exceed $30 \times 30 \times 30$ cm^3. You need to consider that the patient/resident may be walking outdoors or indoors. Usually the nurse walks the patient within a 30 m diameter. In urgent situations the nurse will need the chair to be available within one minute.

11.11 DISPOSABLE BLOOD PUMP

In blood treatment applications there is a need for a small disposable pump to reduce the interaction of the patient blood with operators and equipment. The pump can be placed along with other disposable fixtures. You are required to design and build a pump with the flowing specifications:

- All blood-contacting parts are disposable.
- The pump can be mass produced to reduce cost.
- The pump must be able to achieve a variable flow rate from 0 to 100 m/min with a volumetric accuracy of $\pm 5\%$.

- The pressure head is between -100 mm Hg to 100 mm Hg.
- Disposable size is not to exceed $30 \times 30 \times 25$ cm^3. The overall fixture must not to exceed $30 \times 30 \times 30$ cm^3.

11.12 NEWSPAPER VENDING MACHINE

The specific problem with existing newspaper dispensers is the ease of stealing the papers in the machine when the door is opened. Theft from existing newspaper vending machines is a source of revenue loss as well as frustration for newspaper companies today. Another limitation of the current dispensers is that they can only handle one newspaper for each dispenser, which makes them inefficient spacewise. You are required to design and build a newspaper dispenser with the following specifications:

- Can handle at least three different newspapers.
- Dispenses one newspaper at a time based on customer selection.
- Newspaper companies make most of their profit from advertisements and not from newspaper sales, so they will be unlikely to purchase machines that cost more than the current ones.
- Familiarity of the design to customers is crucial, since customers (newspaper companies) will be unlikely to purchase newspapers from designs that deviate from existing designs.
- The design size should conform to current newspaper dispenser standards, especially with regard to height.

11.13 PEACE CORPS GROUP PROJECTS

The class is assigned to a Peace Corps mission to village of an average population of 250 somewhere in the universe where electricity is not available. The village needs the team's help to design mechanical instruments that will help the village to function.

11.13.1 Projects

The villagers require help from the team for the following:

- A mechanism to pump water. Their current technology is an open well with a bucket. Because of health reasons, it is best to cover the well and design a mechanical pump. The power for the pump could be provided manually or animal powered. Pumps are needed for both the drinking and irrigation system. Because of limitations in material resources, the pump must be used for both irrigation and for drinking water. (Assume the water is drinkable.) The flow rate needs to be controllable. Drinking water demand is much less than that of the irrigation system (about 1/200). The well is about 100 m deep (Don't ask me how they reached that depth.)

- A mechanism to grind wheat to produce flour. The current technology is that each household has their own manual grinder composed of two rough, heavy disks. The disks are aligned around a central hole where the wheat is fed. This current technology is time consuming and a major waste of resources. It is best to set up a central location for this task that every household could use. The new mechanism must save time and could be a one-man operation.

- A mechanism to help in seeding agricultural land. The current technology utilizes a shuffle and digger. If you have ever planted seeds by hand, you will know that this is a labor-intensive job. Average land per household is about 1200 square meters (30 households). The system needed should be energy independent. Please do not humiliate the people by suggesting a simple attachment to a horse.

- It is assumed that with the technology developed to pump the water and seed the ground, the yield will become 10 times higher than before. The task now is to develop a mechanism to collect the wheat and vegetables. Two systems are needed:
  (a) One for wheat collection and packaging; and
  (b) Another for vegetable collection and packaging.
  The average temperature is about 27°C during a year.

- A system to generate electrical energy would be of interest to the village. It is very clear one source will not be sufficient. They have a lot of wood at their disposal. Wind is more viable than solar because of material.

- A time and calendar system viewable by the whole village. Currently time is known by the shadow length of a well-designed wood column. What is needed is a mechanism that tells the time, day, month and year.

11.13.2 Materials

The village is out of touch when it comes to the availability of material. You are allowed to have $100 worth of material shipped to you. Be wise in using the material. Remember, lumber is available by the tons.

11.13.3 Machining

Machining would be the greatest challenge. There is no machine shop available. What is available is manual driven wood shaping equipment.

11.13.4 Deliverables

1. Teams will show three different alternatives of their designs using Pro/E or another CAD tool: Teams will apply a systematic design process in reaching three alternative designs using CAD tools. Teams are encouraged to be creative and not restrict themselves to ideas that are currently available.

2. Teams will show the analysis of the three alternatives as part of the evaluation process. In addition to the design objectives, the analysis is an important part in reaching a design decision. Details of the analysis are required.

Index